KURTIS KRAFT MASTERWORKS OF

and already one of the national stars of AAA Championship racing. Kurtis' design for Rex was an instant success. Kurtis felt that if a car could be built with a stiff frame, it could be given softer springs, which would give a better ride and better handling. In a day when race cars had very stiff leaf springs it was an almost revolutionary idea. In competition the idea proved sound, and Roy Russing drove Rex's car to two West Coast championships. Kurtis did a brisk business repairing cars damaged at Gilmore and the other local midget tracks, and he worked on crash-damaged big cars for Lou Meyer, Billy DeVore, and Gus Schrader.

By the end of 1938, Kurtis tired of the arrangement with Thorne and Adams and moved out, setting up his own shop again behind the top shop on San Pedro Street. Race drivers and little boys hung around the shop watching the cars go together. One of the neighborhood youngsters was named Johnnie Parsons, a boy who as a driver later did quite well at Indianapolis with a Kurtis car. At about the same time, Clyde Jones built the first midget with a headrest. Although it did not make the cars go any faster, suddenly everyone needed a headrest, and Kurtis was kept busy updating other midgets with that latest stylistic necessity.

The success of Rex Mays' car brought in more business, and Kurtis at one time had six midgets under construction. Buyers included Bob Swanson, Jack Prickett, Ted Halibrand, Roy Sherman, Willie Utzman, Mel Hansen, and Perry Grimm, all well-known midget drivers or owners. Several Los Angeles area builders of Indianapolis and dirt track cars also turned to the Kurtis shop to help them build bodies and chassis parts.

By 1939 Frank Kurtis had another itch to see the world outside Los Angeles. Some of the midget drivers for whom Kurtis built cars had graduated to the big time at Indianapolis, and Kurtis had already done some body and frame work on Indy cars for Kelly Petillo and Fred Frame. He decided it was time to take a look at the Speedway. Frank, Arlene, and Arlen piled into the family Buick and headed for Indianapolis.

Kurtis had arranged with Charlie Allen to get a pass to the Indianapolis garage area. Allen was part of the team working on the W.B.W, an Offenhauser-powered car belonging to Ed Walsh, scion of a wealthy St. Louis family. The car, a 1932 Miller originally driven by Gus Schrader, had gone through several hands and needed a new fuel tank. Allen enlisted Kurtis to do the metal work on a rush basis before driver Floyd Davis could attempt to qualify. Kurtis, with 20 years of metal working experience, finished the job in a couple of days.

Leon Duray's *Duray-Barbasol Special* Number 26 had been run into an infield creek by Ronnie Householder, injuring him painfully but not critically, and the car needed major body work. Duray asked Kurtis, ex-riding mechanic Johnny Pawl, and Reynold MacDonald to

After Kurtis left the Thorne shop, he returned to South San Pedro, this time at the Pioneer Top Shop at Number 6215. The sign on the window reads "Race Cars Custom Bodies by Frank Kurtis." The shop also worked on an autogiro, seen through the passageway between the buildings. *Arlen Kurtis collection*

This Lloyd Axel rail-frame Kurtis midget was built with an Offenhauser engine. This photo shows it as it appeared after restoration. *LeRoy Byers*

The 1939 Leon-Duray *Barbasol Special*, after Ronnie Householder rolled it into a ditch at Indianapolis. Frank Kurtis, Johnny Pawl, and Reynold MacDonald repaired the car in time for Billy Devore to qualify it and bring it in 10th in the 500. *Eastern Museum of Motor Racing*

In 1940 Marty Keller was getting the Bill Hughes Indianapolis car ready for the Speedway, and Kurtis worked on the car for a week in Marty's machine shop in Denver. The car was one of the Miller-Fords of 1935. Harry Syer, one of Hughes' employees, recalled that Frank built a set of headers for the 255 Offenhauser engine out of flat steel stock. Frank is seen in this photo standing behind the Number 17 car in pit lane at Indianapolis. The Hughes car's driver was George Robson, in the white coveralls. Robson would win the 500 in 1946 in a Joel Thorne car that Kurtis helped build. The Hughes car finished 23rd in 1940, with suspension damage. *Bruce Craig*

rebuild the car in time for qualifying, 12 days away. Kurtis rented an empty storefront on North Meridian Street as a garage and they labored in relays, 24 hours a day, to get the car back together.

Duray's car was an odd chassis, put together by Ernie Weill, a former employee of both Harry Miller and Fred Offenhauser, and powered by a special 182-cubic-inch four-cylinder supercharged Miller engine. Weill, an immigrant from Germany who had been the court jester of the Miller plant, oversaw the repairs, cracking up Pawl, Kurtis, and MacDonald with his jokes while they labored over the bent and twisted sheet metal. After a series of all-nighters, they got the car back to the track on the last day of qualifications.

Young Billy DeVore, in his third year at Indianapolis, was hired to replace the injured Householder. DeVore qualified the Duray car at 116.527 miles an hour, good enough for the 33rd starting spot. DeVore charged through much of the field to finish 10th, though his car was nearly 9 miles an hour slower than that of winner Wilbur Shaw.

The Walsh *W.B.W.* car eventually qualified at 119.375 miles per hour. It broke a shock absorber and made two pit stops for repairs. Finally, on his 43rd lap, Davis had to quit the race. He finished 27th.

At the end of the summer of 1939, the German blitzkrieg swept into Poland, beginning the war that would

HARRY KISTLE
SALES
KELLY
TIRES
KELLY TIRES

previous page
Bill Hughes asked Kurtis to build a sports car for him around the Ford V-8 racing engine he took out of the 1935 Miller-Ford chassis. This boat-tailed speedster was the result. *Arlen Kurtis collection*

soon engulf America. Racing continued on borrowed time. Kurtis was still customizing street cars and building trailers and trucks, but as the war overseas progressed, the race car shops in Southern California were more and more swept up in defense preparations. In May 1940, President Franklin Roosevelt issued a call for a 50,000-plane air force, a proposal that brought a boon to the local aircraft industry. Plants such as Fred Offenhauser's race engine shop began to do subcontracting for Lockheed and North American Aviation. Thorne's shop was converted to war work, building landing gear for Lockheed Aircraft.

While the war clouds gathered, Kurtis became a regular at Indianapolis. In 1940 he headed east with Jack Prickett and his midget, running in races along the way. In Denver, Kurtis worked on a car owned by Denver Ford dealer Bill Hughes, helping Marty Keller install a new 255 Offenhauser engine in Hughes' 1935 Miller-Ford chassis Number 17, which George Robson was to drive.

When he reached Indianapolis, Kurtis was again pressed into service repairing broken and wrecked race cars. He would say later, "I sure was getting experience."

Robson qualified the Hughes car 23rd at 122.562 and finished 23rd when a shock absorber failed on Lap 67. Later Hughes asked Kurtis to pull the Offy out of the Miller-Ford, put a Ford V-8 back in it, and set it up for road use.

That summer, after Indianapolis, Kurtis also worked on Joe Lucas' and Lou Fageol's Gold Cup boats, and built a batch of junior midgets, a child's toy race car similar to those that became known after the war as quarter midgets.

Ed Walsh was also back at Indy in 1940 with an old Harry Miller chassis, the ex–Gus Schrader four-wheel-drive car that Lou Moore had rebuilt with an odd front suspension. The exhaust pipe fell off and the car finished 16th for driver Louis Tomei. Walsh, not happy

THE FIRST KURTIS-KRAFT

In 1940 Frank Kurtis worked on the Indianapolis pit crew of George Robson's car, a Miller-Ford chassis then carrying a 255 Offenhauser engine. The car and engine were relics from one of Harry A. Miller's later projects, the Miller-Ford Indianapolis racing cars. Urged on by Preston Tucker, later to become famous as the promoter of the post–World War II Tucker Torpedo, Miller persuaded Henry Ford to enter a 10-car team at Indianapolis in 1935.

There was far too little time to complete the cars and test them in the two months available, though Miller and his Ford employees worked day and night in the effort. Only 4 of the cars qualified and on race day, though they were reasonably fast, all 4 fell out with steering problems. The front-drive, independent suspension chassis were quite innovative for the mid-1930s, but an embarrassed Ford put them away after the race. Most of them reappeared in later years with a variety of engines. The 1941 Novi is the best known of the revamped Miller-Fords, and Andy Granatelli's 1948 Indy entry remained in competition the longest of the 10.

The original Miller-Ford V-8 engines, hopped up by Ford, had high-compression aluminum heads, high-lift Winfield cams, and four Winfield carburetors on a special manifold. Billy Hughes of Denver bought one of the cars in 1939, had Kurtis narrow it from its original two-man configuration to a single-seater, stuffed an Offy in it, and went racing at Indianapolis. With an extra engine on his hands, Hughes asked Frank Kurtis to build a sports car around the Indy Ford V-8.

Working in his South San Pedro Boulevard shop, Kurtis took a 1934 Ford frame and shortened it, adding 1939 Ford hydraulic brakes. He built a framework of steel angle iron material, over which he formed a hand-made body, and attached modified Chevrolet fenders. The top folded into a recess behind the seat. The car had no trunk, and the spare tire slid into a storage area accessible from inside the car.

Kurtis told interviewers later that he charged Hughes $800 for the job. The car was the first to carry the Kurtis-Kraft logo.

Bill Kenz detuned the engine for street use, and Hughes drove it throughout World War II. In 1947 Hughes returned the car to Los Angeles and *Popular Mechanics* did a cover story on it. Mrs. Richard Winton then bought the car and drove it for a time in California. She and her husband subsequently moved to Tangiers, Morocco, and took the car with them. The Wintons sold the car in Morocco, and the Kurtis-Kraft went through several hands until it needed repairs that the owner could not afford. Left in storage, it surfaced in 1975 in the hands of a French schoolteacher and car collector.

Impressed with the car but unaware of just what it was, the Frenchman, Jean-Pierre Leclercq, repaired the engine and undertook a search for the origins of the chassis. Leclercq traced it back to Frank Kurtis, still living in Glendale in 1979.

The 40-year-old Kurtis custom was still driveable, once the Mercury V-8 engine was freed up, so Leclercq drove it from Casablanca to the ferry at the Straits of Gibraltar, up through Spain, across the Pyrenees mountains, and through southern France to Aachen, Germany. He showed it at car meets in Germany, finally driving it to Normandy, on the coast of France. The original Indy Ford engine was long gone and so were the Moon hubcaps, most of the instruments, and various other parts, but Leclercq went on to undertake a frame-off restoration of the car.

with his car's performance, hired Kurtis to go to St. Louis that summer and rebuild the car. (Moore bought the chassis from Frank and Al Scully, who got it in Harry Miller's bankruptcy sale. The 308-ci V-8 engine in the FWD car, destroked to 269 cubic inches, went to Bunny Phillips, who put it into a type 35 Bugatti chassis.) Kurtis lived for weeks in a loft room in the Walsh home while he worked on the car. He used the engine and running gear from the old Miller, but from photos that show the frames of both cars, it is clear that he built a new chassis.

Kurtis' effort was an entirely conventional car, if a somewhat more attractive one than some of the junk formula racers of the 1930s. The 1941 Indianapolis entry form for the Walsh car listed Kurtis as the builder. It was his first Indianapolis racing car. Entered as the "*Tom Joyce 7-Up Special* Number 28," Sam Hanks qualified it 33rd, but the engine blew in a test run the day before the race. The car crashed and was damaged too badly for even Kurtis to fix overnight.

On race day of 1941 one of the Gulf-Miller rear engined cars caught fire, and the garages on the south side of Gasoline Alley burned. Kurtis had left his tools and a movie camera in the Walsh garage and lost all of them in the blaze. That fall Walsh asked Kurtis to pick up the wrecked 7-Up car from storage in St. Louis and rebuild it once more. Frank had just gotten it to Los Angeles and started to straighten it out when the Japanese bombed Pearl Harbor, and World War II put racing on hold for the duration.

Kurtis built more than a dozen junior midgets for youngsters between 1939 and 1941. This one was driven by young Don Grosh, one of the participants in weekend races at the 1/10-mile track Johnny McDowell operated for bus company mogul Lou Fageol. *Arlen Kurtis collection*

After the disaster at Indianapolis, Kurtis returned to the Thorne shop in Burbank. Art Sparks was growing ever more disenchanted with Thorne, and Joel asked Kurtis to take over managing the plant at $3,000 a year plus 50 percent of the profits. As favorable as the deal sounded, Kurtis said, Thorne meddled in the contracts and made it hard for him to manage the shop. His employees there included race driver Chet Miller, who would later drive a Kurtis-built Novi at Indianapolis, and Ed and Zeke Justice, who later would become key employees in Kurtis' own shop.

While waiting for Thorne's defense business to get into high gear, Kurtis built a few special projects,

In 1940, after parts fell off the old Jack Holley car during the race, Ed Walsh asked Kurtis to go to St. Louis and build him a new car for Indianapolis. Frank built this car, using the Offenhauser engine and a few parts from the eight-year-old Miller. The Walsh Kurtis-Offenhauser is seen here with George Connor at the wheel at Indianapolis in 1946, when it finished 24th with a burned piston. Behind the Walsh car is the *Ross Page Special* Number 41 that Kurtis built, driven by Mel Hansen, and the *Erwin Wolfe* Number 7 of Frank Wearne, the same car with which Wilbur Shaw won the 1937 500. *Edy Kurtis collection*

During World War II, the movie studios begged Kurtis to build replicas of the Army's jeep, a vehicle too valuable to be spared as a movie prop. Frank put this version together out of Ford parts. Presumably he replaced the 1928 and 1935 Ford wire wheels with more appropriate parts before filming began. *Watkins Glen Museum*

including a copy of a U.S. Army jeep for the movie studios and a three-wheeled, rear engine car powered by a Ford V8-60 engine.

The first model of the three-wheeler was steered by the single rear wheel. Kurtis took the prototype out for a test run in traffic and quickly found that the back end swung out of its lane before the front wheels began to turn. Kurtis said later he was sitting in the shop, disgusted, when a P-38 fighter swooped low overhead, landing at the nearby Lockheed air field. The tricycle gear of the P-38 inspired him to rebuild the car, putting the single wheel in front. Thorne later crashed the car, killing a girl who was riding with him. Kurtis rebuilt the vehicle and after the war, Thorne sold it to promoter Gary Davis.

Materials shortages and chaotic allocations and priorities assigned by the War Production Board, compounded by Thorne's meddling, made it difficult for the plant to deliver on its contracts. Kurtis, tired of dealing with Thorne's unpredictability, quit.

Kurtis set up his own wartime shop on Brand Boulevard in Glendale, where with his father he struggled

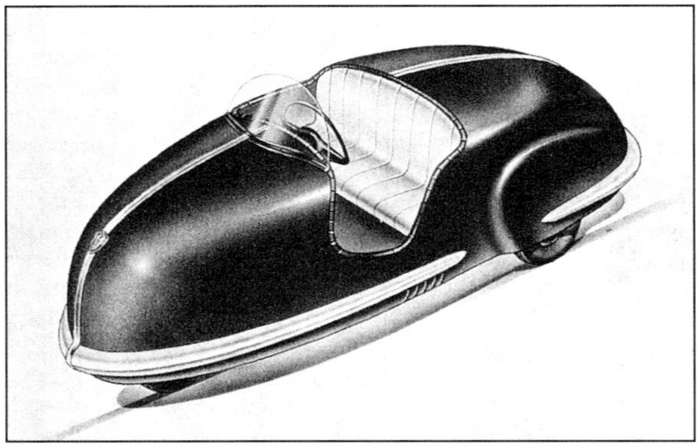

Kurtis built a three-wheel car out of Ford parts in 1942, while working part-time for Joel Thorne's shop. The idea was to save on rubber, already scarce once the Japanese had captured the rubber plantations in Java. The first version had the single steered wheel in back, but it did not handle well, so Kurtis rebuilt it with the steerable wheel in front. After the war Thorne sold the car and its plans to Gary Davis, who promoted it shamelessly as the Davis Californian, but only prototypes were built. *Arlen Kurtis collection*

to make ends meet. They built tooling for Plexiglas aircraft windows, domes, and canopies for Lockheed aircraft, North American P-51 Mustang fighters, Northrop P-61 Black Widow night fighters, and Boeing B-19 and B-17 bombers.

Thorne and Art Sparks went to Washington and won approval to set up another plant in Phoenix, Arizona, to do war work, but after 18 unproductive months, the Navy canceled its contracts. Some of Kurtis' own wartime work came in the form of subcontracts from Thorne. Another wartime Kurtis project was building molds for the large Swedlow Plastics company. The exposure to the aircraft industry and its techniques taught Kurtis how to weld aircraft-quality chrome molybdenum steel tubing and how to shape aluminum forms inexpensively, in production quantities, as well as the value of quick-release Dzus fasteners and other aircraft hardware.

But without specialized machine tools and the priorities or the money to buy them, Kurtis faced difficult times during World War II. There were periods when contracts from North American or Lockheed would keep the shop busy, but not on a full-time basis. Unlike Fred Offenhauser, whose shop made a nearly $100,000 profit in 1942 alone, Kurtis struggled.

Mert Harris, Kurtis' friend from Michigan, worked for the Ford Motor Company during the war as an engine expert on R-2800 aircraft engines. Harris recalled visiting Kurtis while he was on troubleshooting trips to Northrop Aircraft in Hawthorne, California, in 1943 and 1944. Harris stopped by the Brand Boulevard shop several times and found the Kurtis workers building children's toys, using wood scrap and other nonstrategic materials. Before Christmas 1943, Kurtis built 5,000 child's scooters, using women, teenagers, and other nonessential workers. A year later, Sears Roebuck & Company asked Kurtis to build a wooden toy for the Christmas season, and he designed a little car about 14 inches long with a small mahogany boat and trailer. Sears wanted 5,000 of the toys, far beyond what Kurtis' shop could turn out in such a short time. Kurtis later drew up a toy model racing car, but when the war wound down he sold the rights and tooling for $300.

In 1944, with time on his hands, Kurtis built a smooth little pedal car for Arlen. Its lines would show up a year later in larger form as the Kurtis-Kraft midget.

World War II was costly in many ways for the Kurtis family. Frank's brother, Andrew, volunteered for the

Army's 82nd Airborne Division. In October 1943, Andrew was killed during a parachute jump at Fort Benning, Georgia.

Another of Kurtis' brothers, Bill, went into the Navy and served in a gun crew aboard the minesweeper *YMS-100*. Bill was temporarily deafened in a 3-inch gun incident in the Pacific in 1942, and he never fully regained his hearing.

Two days after Andrew's death, Arlene Kurtis gave birth to a daughter, Ellona, at home in San Pedro.

In the fall of 1944 defense contracts began to be canceled, and there were hopes that racing would resume for the 1945 season. Drivers who were on the West Coast began to drop in on Kurtis and talk about new midgets and new cars for Indianapolis. Then the Germans counterattacked in the offensive that became known as the Battle of the Bulge, and the Japanese unleashed their deadly Kamikaze attacks. The war, it seemed, would go on for a while. There would be no racing in the first months of 1945.

Despite the renewed fighting, it was obvious that the war would be over soon. Germany finally surrendered in May 1945, and although a million-man invasion of Japan seemed possible for the late summer, the end was in sight. The pent-up interest in racing was about to blossom into a new burst of enthusiasm that would carry Frank Kurtis to the height of his career.

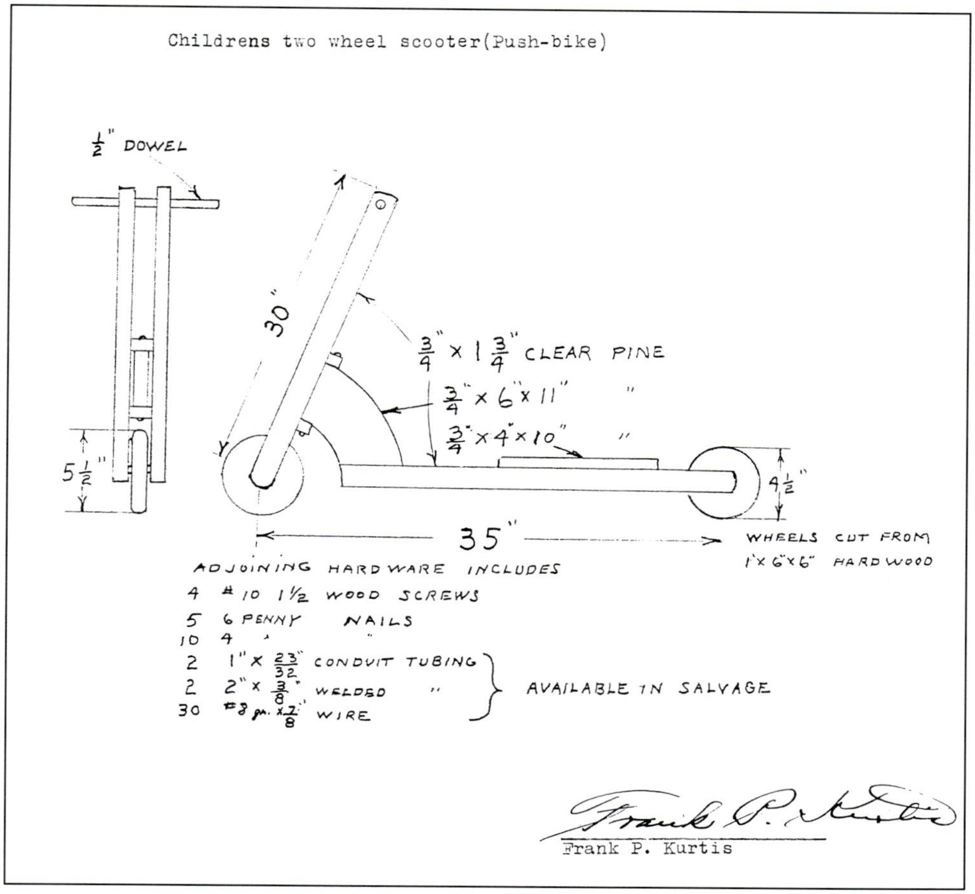

During the war, to help keep the shop busy when defense contracts lagged, Kurtis designed this child's scooter, built out of scrap and nonessential materials. *Arlen Kurtis collection*

In 1944, Kurtis drew up a design for this child's pedal car for his 11-year-old son, Arlen. The smooth lines of the little car would be reflected later in the full-size postwar Kurtis midget.

The Postwar World

CHAPTER 3

KURTIS MIDGETS

Frank Kurtis said of his postwar midget design, "The shape of the body I drew simply to cover the radiator, cowl and fuel tank with a pleasing line. I had the hood extend back over the cowl for ease in working on the instruments. At the same time, I made the flare for the cowl as a separate plastic windshield. The headrest made the driver appear lower in the car." Years later he said that midget was his greatest pride.

Shortly after the turn of the new year in 1945, Kurtis sat down at the Tam O'Shanter Inn in Glendale with his old racing buddies, Buzz Lowe, Dean Whalen, and Tom Lewis, and sketched out on the back of a place mat his idea of a postwar midget. With sweeping lines, he drew a sleek tail with a headrest. The car would have a frame of aircraft-fashion 4130 chrome-molybdenum tubing and torsion bar springs. It would be covered with an envelope body.

The little car that emerged from the rough sketch was smooth and streamlined, far more modern looking than the prewar midgets. "That's what I'm going to build," Kurtis said, then crumpled up the mat and stuffed it in his pocket.

The drawing Kurtis made in early 1945 would become the most attractive racing car of the mid–twentieth century. It set the style for racing cars of all sizes in the United States and influenced design abroad. Not since Harry A. Miller's long, low front-wheel-drive cars of the mid-1920s had such a design so well captured the sense of its time. With World War II coming to an end, the world was looking forward to new things. Kurtis and dozens of industrial designers stepped up to move beyond the art deco world of the 1930s into the new and optimistic postwar era.

Drawing on his wartime experience with aircraft materials and his belief that chassis stiffness would allow the use of softer springs to give better traction, Kurtis laid out a midget chassis that was far more rigid than the channel-section frame rails of then-conventional

Eight midgets, all Kurtis-Krafts, in an AAA 100-lap championship race at Williams Grove Speedway September 13, 1953. Len Duncan in the Number 19 is out front, followed by Joe Barzda in the Number 4, Ed Johnson in the Number 21, Bill Homeier in the Number 34, with Jiggs Peters in the Number 5, and Cal Niday in the Number 2. *Author's collection*

Bill Schindler in the most famous eastern midget, the Mike Caruso Number 2, seen here at the board track in the Polo Grounds in New York City in 1948. Schindler would win 53 feature races in the Caruso cars in both 1947 and 1948. *Author's collection*

Eddie Jackson with the former Lloyd Axel Number 5 Kurtis Offy midget, with the body off. Jackson won two straight Rocky Mountain championships in this car, the third Kurtis to be built after World War II. Note the dry-sump oil tank under the seat, the Pesco aircraft fuel pump, and the Kurtis reinforcement on the seat. Jackson is wearing one of the Hook-Fast belt buckles all good racing people had in those days. *LeRoy Byers*

racing cars. As Kurtis said in a 1980 letter to the author, he had studied the use of the torsion bar springs on Wilbur Shaw's 8CTF Maserati and thought they could be adapted to his stiff tube-frame design. Mercedes and Auto Union had used tube frames in Europe before the war, but American builders had not adopted the idea.

Rex Mays had been very happy with the prewar midget Frank had built for him, and his endorsement of Kurtis' cars had helped get Frank started in the midget business. Excited by Kurtis' ideas for postwar cars, Mays encouraged him to start building his new design midget as soon as he could. They also discussed a new chassis for the Bowes car Mays expected to drive at Indianapolis when the track reopened.

Even while the fighting was going on, Kurtis was readying his shop for the end of hostilities. He bought a larger building at 1107 East Colorado Boulevard in Glendale, and in his spare time began laying out tools and fixtures for the midget's frame, front axle, and other parts. When Germany finally surrendered and the Allies moved closer to victory in the Pacific, Kurtis found aircraft subcontractors eager for peacetime work. Expecting to build as many as 30 cars a year, he had drop-forming dies made for the hood, fuel tank, and half-sections of the tail and nose of the new car, and ordered dies and mandrels to form axle ends and other parts. Kurtis began to assemble a work force from among veterans of the aircraft plants, many of whom had learned the necessary skill of fine quality welding during the war.

By the summer of 1945, with a few racing people beginning to filter home from Europe, word spread about the planned new Kurtis midget with its tube frame and other innovations. Ten buyers stepped up to put deposits on cars.

While his defense contracts were being canceled, Kurtis constructed a jig in which midget frames would be built. He was ready to start bending metal for the first car well before the Japanese surrendered in August, but materials were a problem and shortages kept him from starting work.

In order to get materials to build the first prototype midget Kurtis turned to Leland "Buzz" Lowe, an old friend who was running a machine shop doing work for Bendix Aviation. The Bendix purchasing agent happened to be in the shop one day when Kurtis and Lowe were talking about the new midget. He offered to get around the materials problem by ordering the aircraft tubing and sheet aluminum they needed for the Lowe shop under a Bendix priority. When the shipment showed up, Kurtis took a trailer to the Bendix loading dock, picked it up, and went his way. It provided enough metal to build two cars, but Lowe had no rear axle. Until early 1946 when rear axles and steering units became available, Lowe could not take delivery of a car for himself.

Not only aircraft-quality steel tubing and sheet aluminum remained in short supply. The companies—Bennett, Norden, and Conze—who made rear axle assemblies, steering gears, and hubs had not yet gotten back into production. Their parts required aluminum and magnesium ingots and foundry and machine work. In the absence of better components, the prototype Kurtis-Kraft midget was built with a steering gear converted from a Franklin passenger car and a prewar Bennett rear axle.

UPRIGHT POSTWAR MIDGETS

The post–World War II Kurtis-Kraft midget was the first purpose-built racing car designed for quantity production on an assembly line, though the Kurtis line hardly approached Detroit standards. Kurtis had molds made for pressing complex aluminum shapes, and achieved some degree of division of labor in his shop.

Frames were made of 4130 chrome molybdenum aircraft-quality tubing, welded with mild-steel rod. The body panels were supported on hoops of smaller-diameter tubing and attached to small brackets that accepted Dzus 1/4-turn fasteners. Tube, for its weight, is far more stiff than the channel-shaped frame rails long used on vehicles of all types. The body framing added to the stiffness of the chassis, though at first the placement of the truss structure was not designed to take full advantage of the technique.

By adding stiffness, Kurtis controlled the geometry of the suspension much better, allowing softer spring rates, which in turn permitted the wheels to follow the track surface better, adding to the available traction. To understand the improvement, consider an extreme example: the wheel hop of a passenger car on a corrugated gravel road. The closer the wheels follow the surface the better the traction and the faster a car can negotiate a turn before centrifugal force pulls it out of control.

The Kurtis design also reduced unsprung weight, which also permitted the wheels to follow the surface better. This was achieved by making the entire spring "sprung weight," unlike the previously used leaf springs, as well as by using the new, light Hills-McCanna (later Halibrand) magnesium wheels. At Indianapolis, where such things could be measured closely, the mag wheels were lighter and stiffer than the previously used wire wheels enough that speeds went up by 2 miles an hour.

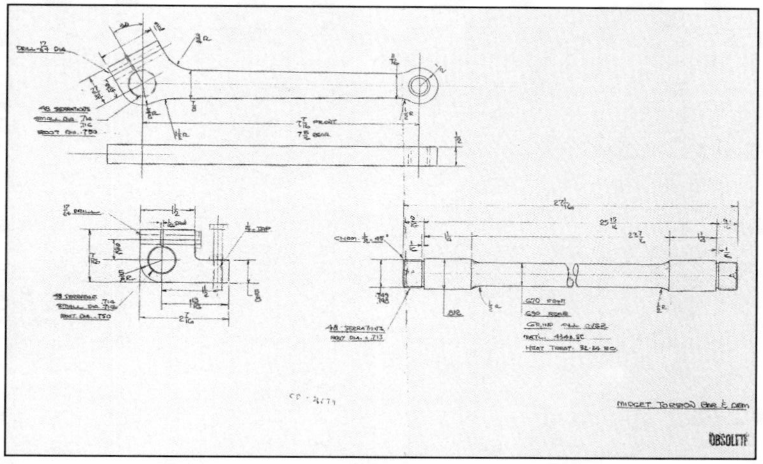

The torsion bars and arms for the Kurtis midget. Leo Goossen, who had drawn the plans for virtually every American racing car since the 1921 Millers, did this tracing in July 1947. *Leo Goossen Archive*

With orders for 10 midgets already in hand, Kurtis began work at the shop in Glendale. That fall, while the midgets were starting to go together on one side of the shop, on the other side Kurtis was rebuilding the old 1939 Duray Indianapolis car for Ross Page, and working on Ed Walsh's Indy car. Kurtis offset the engine of the Ross Page car 2 inches to the left, allowing the driver to sit low alongside the driveshaft, a design that, seven years later, would be called a roadster.

V-J day came with the suddenness of the atomic explosions over Japan, but racing people were ready. From coast to coast, cars came out of mothballs. At Freeport, New York, and a score of other tracks, Labor Day weekend 1945 saw the resumption of full-bore midget racing, even though the Office of Defense Transportation ban was not officially lifted until September 16. Freeport's first crowd numbered 15,000 people, a huge one even by comparison with the best of the prewar racing.

At Williams Grove, Pennsylvania, sprint cars raced on September 9, attracting 34,200. In Los Angeles the track at the Bonelli Ranch opened postwar midget racing in California before a crowd of 18,000. The long-delayed Coliseum races went forward on October 11, 1945, attracting 45,000 paying fans to the new asphalt track "Hollywood" Bill White had laid down. Both he and officials

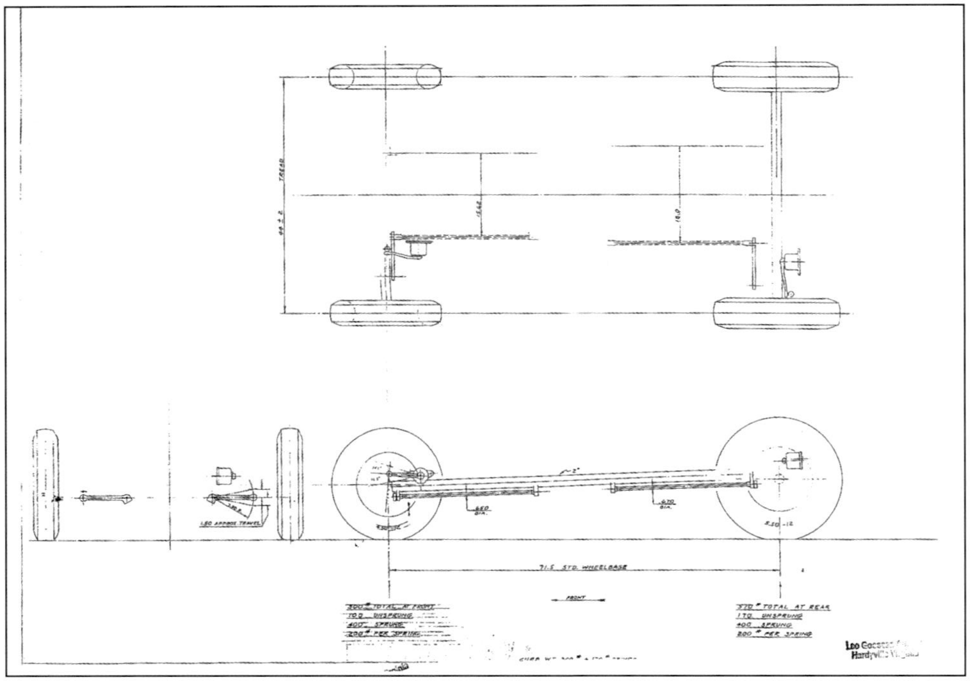

This is John Bond's drawing of the basic frame of the Kurtis-Kraft midget, done more than a year after the first cars were produced, to tidy up Kurtis' own rough drawing.
Leo Goossen Archive

Previous page
In the Kurtis shop on Colorado Boulevard in late 1945, the midget assembly line crowded the Indianapolis cars into a corner. Seen here are five midgets. The car second from right with an Offenhauser engine installed is Roy Russing's midget, the first one to be completed. The large car in the foreground is the 1940 Walsh car, which would run again at Indianapolis in 1946 and 1947. This photo, showing the typical Kurtis frame construction, leaves no doubt that the car was not just a rebuild of the 1932 FWD Miller that Walsh had owned. Kurtis himself is seen at the far left, showing Ed Haddad the midget under construction for him. *Watkins Glen Motor Racing collection*

of the United Racing Association were stunned by the turnout. While midget purses of a few hundred dollars had been standard before the war, this one was $15,000. In noninflated greenbacks this was equal, in 2001 dollars, to probably $100,000. Auto racing had been the Number One spectator sport in the nation before the war, and it seemed certain to keep that ranking in the immediate postwar years.

Kurtis' younger brother, Bill, was discharged from the navy that October and took over as business manager of the Kurtis company, then known as Kurtis-Kraft, leaving Frank free to build the midgets and Indy cars then on order. It quickly became apparent that the Colorado Boulevard plant was too small. In mid-1946 Frank sold that shop at a profit of $16,500 and bought a 15,000-square-foot building on North Alger Street, just across the city line in Los Angeles, backed up against the Southern Pacific railroad tracks.

Roy Russing, one of the people who ordered an early Kurtis-Kraft midget, was able to find two Bennett rear axles and a Franklin steering gear, so he got the first car, delivered in December 1945. Materials shortages continued to hamper production, but nine more cars were completed over the next four months. May's Number 44 midget was flown to him at Indianapolis in May.

Vast quantities of aircraft sheet aluminum, fittings, instruments, and other hardware soon became available as war surplus in the Los Angeles area. Kurtis, having worked for North American, Northrop, Boeing, and Douglas Aviation, said with some understatement that he "had a chance to visit their surplus departments."

Kurtis' adaptation of aircraft techniques to race car construction raised the standard of American racing car technology significantly. The design of the midgets incorporated nice touches like the spring-loaded hood straps, the formed Plexiglas windshield, and other clever fittings. Some of his innovations and adaptations, such as the use of torsion bars and tube frames, were major; others, less visible, were actually quite significant. The origins of three items in particular are of interest.

Roy Russing, seated in his new midget, about to take delivery on the first postwar Kurtis-Kraft. Seen here are the distinguishing features of the first cars off the line: The suspension on the first few cars used quarter-elliptic springs. The side panels ran from the radiator to the midpoint of the seat. Later these would be divided at the cowl line. The body panels were attached with 5/16-24 machine bolts set in grommets. Later the bolts would be replaced by Dzus quick-release fasteners. The hoods on the early cars were flat, while after the first 10, Kurtis raised the cowl by an inch, giving more driver leg room and a more eye-pleasing line. There were also running changes in the construction of the dash and firewall, and the nerf bar on the rear radius rods would be attached at the lower rail rather than at the top. *Bill Kurtis collection*

DZUS BUTTONS

George Dzus, of Islip, New York, designed a quick-release fastener in 1925 that took only a quarter-turn to operate, yet gave a positive lock. This was highly important on panels for aircraft, where flutter and vibration might be intense. The Dzus fastener was first adopted for production use in 1933 on the Martin B-10 bomber. During World War II, the Dzus fastener became standard on most U.S. aircraft.

Zeke Justice, one of Kurtis' employees, sent to buy some surplus hardware from one of the aircraft plants, brought back several pounds of the Dzus aircraft screws. Figuring that quick-release Dzus bolts would make it much easier to work on a race car, the Justice brothers went to work replacing the standard bolts with the aircraft fasteners. Kurtis at first blew up at them for changing his design without permission, but by the next day, he had cooled off, realizing that it was much easier to work on a car that had quick-release hardware. He adopted them as standard on his fourth midget.

True to form in racing, not all of Kurtis' innovations took hold at first. There was some resistance from the first few buyers to attaching the body panels with the quarter-turn Dzus fasteners. Some buyers wanted the threaded bolts.

The first postwar Kurtis-Kraft, built for Roy Russing, had quarter-elliptic springs in front, hardly different from the Miller chassis of the 1920s. But a few daring owners liked Kurtis' torsion bar design, and even the new-fangled Dzus buttons, and found that they worked very well. Drivers quickly found that the torsion bar–sprung cars handled better. The first buyers soon returned to have Kurtis refit their midgets with Dzus fasteners, for which Kurtis charged them $80 per car.

SPHERICAL BEARINGS

While use of spherical bearings may appear to be a small technical point, Kurtis' new suspension arrangement probably would not have been successful with the old Ford-style tie rod end or the straight ball bearings used previously on racing cars. Spherical bearings allow movement in two planes, necessary to the Kurtis torsion bar suspension design.

Known commonly as "Heim joints," after the first company to manufacture them in the United States, spherical bearings were developed by the Germans in the late 1930s. Allied intelligence officers went over the wreck of a Messerschmitt Bf-109 fighter that was shot down over Britain in 1940. When they saw the bearings

The Kurtis-Kraft midget assembly line had expanded to seven cars when this photo was taken in the Glendale plant in early 1946. The cars at right and third from right are being set up for Ford V8-60 engines, while the fourth from right has an Offenhauser installed, and the others appear to be built for Offys. Note the characteristic reinforcement at the top of the seats. These plates were usually stamped "Kurtis-Kraft," while a serial number was stamped into the right frame rail under the steering gear. The steering, front axle, oil tank, and other parts would also get serial numbers before assembly began. *Greg Sharp collection*

they immediately realized their importance in aircraft control and landing gear linkages. The Allies immediately copied them. The Rose Bearing Company in Britain was asked to make versions of the joint, but the largest manufacturer became the Heim Company of Fairfield, Connecticut. Kurtis, drawing upon his wartime experience, appears to have been, in late 1945, the first racing car builder to use Heim bearings, on his midget torsion bar suspensions. In 1949 he was the first to use rod ends with Heim bearings for other suspension points on the car that became the Belanger Number 99.

THE JUBILEE HOSE CLAMP

In 1923, the British company I. Robinson developed a new hose clamp that had a threaded screw that operated on serrations cut into the clamp strip. It sold under the trade name "Jubilee." This was a much better clamp than its predecessors, and it was adopted by Rolls-Royce for use on its automobiles and aircraft engines. Rolls developed the 12-cylinder water-cooled Merlin engine that was used in American P-51 Mustang fighters. The Packard Motor Car Company was selected to build the Rolls-Royce Merlin engine in the United States under license, and the Jubilee hose clamp became a U.S. military specification item (AN-748) on December 4, 1943.

Kurtis, after initially using the old standard "Noc-Out" hose clamp produced by the Wittek Company in Chicago, shifted to the Jubilee tangential worm type clamp for his cars.

KURTIS MIDGETS REACH THE TRACKS

None of the new Kurtis midgets were finished in time to compete in the bobtail 1945 California season, but by the opening races in 1946, a few were ready to race. In Los Angeles the first 10 Kurtis-Kraft midgets

TORSION BARS

A torsion bar is simply another kind of spring. In a leaf spring the leaves resist bending by the weight of the vehicle, and their resilience smooths shocks from the wheels hitting bumps in the road.

The common coil spring is both bent and twisted by road shocks.

With a torsion bar the spring is rigidly fixed to the vehicle at one end while the other end is attached to an arm that moves with the wheel. The movement of the arm twists the bar, i.e., puts it into torsion.

In a paper Frank Kurtis and Avard Ward wrote for *Automotive Industries* magazine for October 15, 1948, the two described the Kurtis midget torsion bar design, the first on an American racing car, though Auto Union and Maserati had used such springs in Europe before World War II.

As described by Kurtis and Ward, the Kurtis race cars used a conventional rigid tubular front axle and a one-piece rear axle with no differential, a so-called "locked" rear end. The torsion bars were placed under and parallel to the frame tubes, with each end held in a bracket. At the inner end, each bar was anchored while at the axle end the bar was located in a bearing that allowed it to twist. An arm between 7 and 8 inches long was attached to the torsion bar at one end and to the axle end at the other. The torsion arm was attached to the axle end by a stud that rode in a spherical bearing that allowed it not only to rotate but to twist slightly without binding; thus, the spherical bearing was key to the entire setup.

The original Kurtis midget torsion bar was 0.75 inch in diameter, as suggested by standard textbook formulae. In practice, that was found to be too stiff. Static loading calculations suggested a 0.425-inch diameter, but in tests the smaller bars proved to deflect too much. By trial and error the standard diameter of the torsion bars on Kurtis midgets was eventually set at 0.562 inch. On heavier cars larger bars were used; in fact, in tuning the suspension to different tracks, owners used bars of varying diameters, both larger and smaller than 0.562 inch.

Wheel travel on the front of the upright Kurtis midget was 2.5 inches, and on the rear, 3.5 inches. Later that clearance was found to be too little on the slightly heavier Kurtis roadster midgets of 1956, particularly on rough dirt tracks.

Midget Torsion Bar Data

Bar Diameter:	0.562 inch
Effective Length, rear	24 inches
Effective Length, front	22.5 inches
Effective Arm Length, rear	8.0 inches
Effective Arm Length, front	7.5 inches
Static Deflection, rear	4.0 inches
Static Deflection, front	3.37 inches
Deflection rate, rear	10 lb / degree
Deflection rate, front	9.2 lb/degree
Stress at 289-pound static load	60,000 psi
Total Windup Angle	52.5 degree
Torque rate	5,000 in-lb per radian

Midget Car Data

Dry weight	802 pounds
Fuel, 12 gallons	94 pounds
Water	16 pounds
Oil	20 pounds
Driver, average	180 pounds
Total	1,112 pounds
Load, each rear bar	289 pounds
Load, each front bar	267 pounds

(52 percent rear, 48 percent front)

The bars themselves were made of SAE bar stock of 1095 steel, an alloy that allowed the ends to be upset forged and machined to 0.812 inch with a bearing at each end and a 48-spline serration fitting the lever arm to eliminate any possibility of the arm slipping on the bar.

One great advantage of the Kurtis torsion bar was the ability to easily vary the weight put on each wheel. By adjusting a clamp on the inner end of the bar both the weight and ride height of the car could be adjusted at each wheel. That ability led eventually to "weight jackers" that allowed drivers to adjust the weight, and therefore the handling, during a race.

were delivered to their new owners during the late winter months and early spring of 1946.

A columnist for *National Speed Sport News* visited the Kurtis shop in mid-February 1946 to look at the new midget being built there. He wrote that the car "bids fair to become the sensation of the decade in chassis design and body style" (*National Speed Sport News*, Feb. 28, 1946). Reporter John Gliebe wrote that Johnny Balch, who dealt in new and used racing cars, received the second new Kurtis, one equipped with four quarter-elliptic springs. (Balch later owned two Lesovsky midgets that won 1947 and 1948 championships.) Those springs actually did not work very well at all, and almost all of the first cars were converted to cross-mounted springs or Kurtis' innovative torsion bar suspension. The torsion bars were readily adjustable for different track conditions and improved handling by several orders of magnitude.

Kurtis ran the first advertisement for his midgets in late February 1946, listing the first 9 as going to Russing, Ed Haddad, Rex Mays, Lloyd Axel, Perry Grimm, Billy Vukovich, Gib Lilly, Walt Faulkner, and Ed Wallace. Buzz Lowe apparently got the last of the first 10 cars built. (The first batch of 10, Kurtis midgets were apparently not given serial numbers, though numbers were stamped on the frame, front axle, seat, steering, and other parts of subsequent cars.) Later Lowe would ask Kurtis to build him a midget without a headrest, the only postwar Kurtis made that way.

Though Ed Haddad ordered one of the first midgets, when the time came to pay for it, Haddad was short of cash. Johnny Balch stepped up and bought the car, then put Haddad in it as the driver for the usual 40 percent of the winnings. Balch bought another Kurtis to be driven by Perry Grimm. On March 17, 1946, at the opening race at Gilmore, Haddad took the trophy dash in the Kurtis and won the first heat race. On the pole for the night's feature, Ed gave up the lead to Mel Hansen on the 10th lap but roared back at the end to nip Hansen for the checker. Grimm, in his Balch Kurtis, won the best-appearing car award.

While Kurtis was working at full speed on the midgets, he was also completing work on the Ross Page car, the first tube-frame Indianapolis car, and repairing

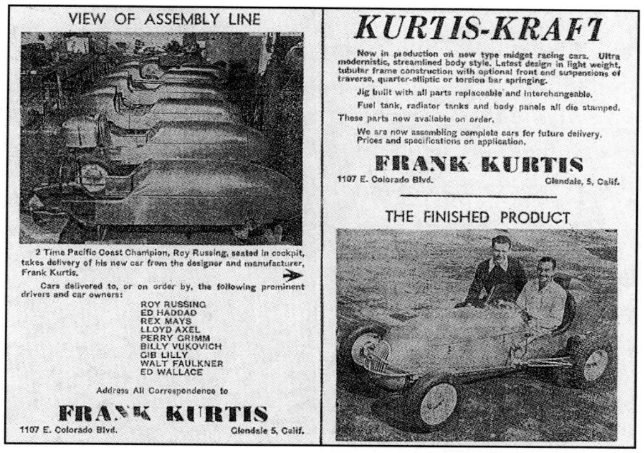

Kurtis-Kraft ran this ad in *National Speed Sport News* on February 28, 1946. Additional ads from the Colorado Boulevard shop ran through early April, when the flood of midget orders made newspaper advertising unnecessary. Kurtis renewed his advertising in *NSSN* in January 1947, from his new and larger plant at 4625 North Alger Street, in nearby Los Angeles. The Haddad, Vukovich, and Faulkner cars were actually built for car owners for whom those men drove. The first new Kurtis-Kraft to make the newspapers as a winner was Lloyd Axel's *Foster Auto Supply Special* Number 5, for a win at Lakeside Speedway in Denver, May 7, 1946. *Author's collection*

Rodger Ward, an eventual two-time Indianapolis winner, seen here in Perry Grimm's Number 65 Ford-powered Kurtis midget, collides with Bill Taylor in Tom Swift's Number 31 Kurtis, while the driver of another Kurtis gets sideways while trying to miss the action at a United Racing Association meet. *Tom Swift collection*

Buzz Lowe in his unique Kurtis-Kraft midget, Number 93 (serial number 121), built without a headrest. Lowe was 31 when this photo was taken in 1948 in Lincoln, Nebraska. Lowe believed his car was the first one built with torsion bar suspension on the front axle. He sold the car at the end of 1948, but bought it back in 1987 and the car remains in the Lowe family. *Lowe Family collection*

Joel Thorne's "Little Six." In April 1946, Kurtis turned to construction of the new Novi chassis, designed at the end of the war by Bud Winfield and Leo Goossen. Though Kurtis could not persuade Winfield to use his tube frame design, the Novi was built with torsion bar front suspension.

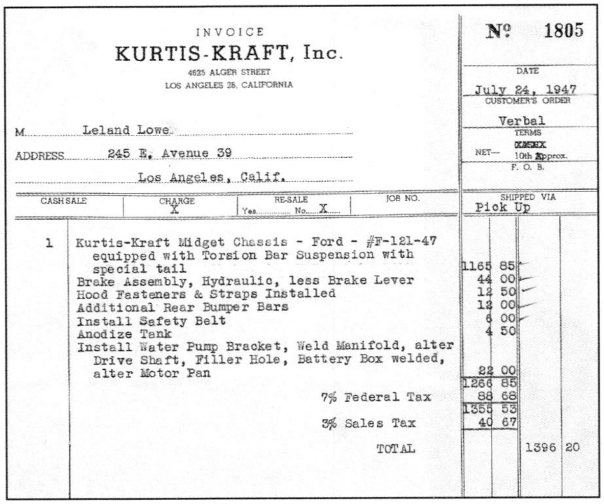

INVOICE
KURTIS-KRAFT, Inc.
4625 ALGER STREET
LOS ANGELES 26, CALIFORNIA

Nº 1805

M. Leland Lowe
ADDRESS 245 E. Avenue 39
Los Angeles, Calif.

DATE: July 24, 1947
CUSTOMER'S ORDER: Verbal
TERMS: 10th Approx.
F. O. B.
CASH SALE | CHARGE X | RE-SALE Yes No. X | JOB NO. | SHIPPED VIA Pick Up

1	Kurtis-Kraft Midget Chassis - Ford - #F-121-47 equipped with Torsion Bar Suspension with special tail	1165 85	
	Brake Assembly, Hydraulic, less Brake Lever	44 00	
	Hood Fasteners & Straps Installed	12 50	
	Additional Rear Bumper Bars	12 00	
	Install Safety Belt	6 00	
	Anodize Tank	4 50	
	Install Water Pump Bracket, Weld Manifold, alter Drive Shaft, Filler Hole, Battery Box welded, alter Motor Pan	22 00	
		1266 85	
	7% Federal Tax	88 68	
		1355 53	
	3% Sales Tax	40 67	
	TOTAL		1396 20

The invoice for Leland "Buzz" Lowe's 1947 Kurtis-Kraft midget, the only one built without a headrest. Note the price of $1,396.20 for a complete race car with Lowe's Ford V8-60 engine installed, needing only paint and a brake lever. *Dean Lowe collection*

Midget racing was having its best year ever and word was quickly getting around that a Kurtis-Kraft was the car to have. Kurtis ran small advertisements in the racing papers during the spring of 1946, but stopped advertising after the Indianapolis race, when the midget order backlog reached six months. As soon as the first Novi and the Ross Page car went out the door for Indianapolis in early May, another 10 midgets were begun. As Kurtis said, "The floodgates opened."

Kurtis-Kraft's earnings grew quickly in 1946, as the production of midgets swelled to almost one a day. In addition to complete cars, Kurtis-Kraft was selling components and spare parts through a network of dealers across the country. Frankie Del Roy's Speed Shop in New Jersey, Johnny Pawl in Indiana, Leader Cards in Milwaukee, and other speed shops carried Kurtis merchandise. Bill Kurtis said the monthly gross was well into five figures, good sales for a relatively small business. From an initial staff of 10, within three years Kurtis-Kraft would grow to employ more than 80 workmen. Shortly after moving the shop to Alger Street, the Kurtises also moved the family home from East 73rd Street in Los Angeles to a larger, more comfortable bungalow on Cordova Street in Glendale.

As demobilization continued in 1946, there was unrest in the nation as a flood of returning servicemen hit the streets. Defense contracts had been abruptly terminated, leaving a gap of months before civilian production and employment came up to speed. A jurisdictional strike between rival unions even disrupted the movie industry, practically next door to the Kurtis plant.

Racing, however, was a way to let off steam for many returning servicemen, and it was making an unheard-of amount of money for drivers and owners. Williams Grove attracted 38,000 fans to its opening 1946 big car meet on April 15, and the pattern was repeated across the country. Freeport, New York, was attracting crowds of 10,000 twice a week in a Municipal Stadium built for high school football.

Top drivers were drawing down $10,000 a year or more. One-legged Bill Schindler, one of the eastern aces, made $20,000 in 1946, more than the salary of many executives.

West Coast events, where Kurtis' midgets were running, soon showed that his chassis was a winner. The money flowing into the sport bought 54 new Offenhauser midget engines in 1945, more than half as many as had been built in the entire eight years of prewar pro-

The Lloyd Axel midget, with its Offenhauser engine, was, in early 1946, the third tube-frame midget Kurtis-Kraft produced. Note the flat hood line, raised by an inch at the rear after the first 10 had been built. *LeRoy Byers*

Vic Edelbrock had two of the best Kurtis-Kraft midgets in Southern California. As a manufacturer of Ford speed equipment, he naturally ran his midgets with flathead Ford V8-60 engines. Despite giving away substantial power to the Offenhauser-powered cars, the Edelbrock midgets did very well even when they ventured to run with the URA Blue Circuit cars. They were both cross-spring chassis, like the Number 27 seen here at Edelbrock's shop at 1200 North Highland Avenue, in Hollywood. In this photo, Vic is making an adjustment under the seat while Don Towle sets a magneto on the SUN machine and Bobby Meeks works on the engine. *Greg Sharp collection*

duction. On the paved tracks, the four–torsion bar Kurtis suspension made the car ride like a Cadillac, and the soft ride kept the wheels glued to the ground. Suddenly even journeymen drivers were wheeling like stars.

A LOOK AT THE COMPETITION

Kurtis, in his success, did not take over the entire midget-building business, though he certainly was threatening to monopolize it in Los Angeles. Before the war, Curly Wetteroth, Clyde Adams, Gordon Schroeder, Bob Pankratz, Roy Richter, and other West Coast constructors turned out a few midgets a year in job-shop style. Now Kurtis was running a race car production line. There were murmurs of discontent over look-alike cars, but the way the Kurtis midgets looked and handled on the track brought customers to his door. While Kurtis was building most of the midgets, Richter turned to a growing parts business for his livelihood. Schroeder, Wetteroth, Eddie Kuzma, and others built a few larger cars for Indianapolis and the half-mile dirt tracks.

The Wheeler Racing Team at Hinchcliffe Stadium, Paterson, New Jersey, 1948. Dutch Schaeffer is sitting in his number 62 car and Tony Bonadies is in Number 21, both Kurtis-Kraft Offys. *Author's collection*

Five midgets, four of them Kurtis-Krafts, led by Eddie Johnson in Mari George's H.O.W. number 21, and Pat Flaherty in the McNamara number 73, in close and furious racing at the Milwaukee Mile in 1955. Note that three of the cars have hiked the left front wheel off the pavement as they round the fourth turn. *Author's Collection*

THE SOLAR AIRCRAFT COMPANY

The Solar Aircraft Company of San Diego, searching for a new peacetime business, started production of its own midget design, an attractive rail frame chassis midget based on a prewar car built by Elmer Ross with a Curly Wetteroth body. Designed with the inexpensive little Ford 60-horsepower V-8 engine in mind, the Ross cars had a large grille opening in front of the radiator, but were otherwise quite sleek. In October 1945, with midget racing doing fabulously well, Solar commissioned construction of a trial run of midget chassis kits, using stainless steel for the frame rails and aluminum for the bodies. The plant where they were built in San Diego was the old Ryan Aircraft factory where the *Spirit of St. Louis* had been put together in 1927.

The first Solar midgets were advertised by Air Associates, Inc. in early 1946. Twelve kits sold almost overnight and Solar began volume production of rolling chassis, less engines. Solar led Kurtis-Kraft as a near mass-production race car operation by a month or two. Its cars were equipped with Norden steering and Bennett rear axles. Air Associates, Inc., handled sales nationwide. The price of a Solar in February 1946, was $670, or $130 less than Kurtis was asking for one of his kits.

The Solar production bodies were tall, with bulges in the hood to clear the V8-60 engine. Their tails were more sharply tapered than the Kurtis midgets, making it easy to tell them apart.

Unfortunately for Solar, Frank Kurtis' midget was a major leap forward in design, while the Solar was obsolete 1938 technology. With cross-spring suspension, it rode hard and could not keep up with the Kurtis midgets. The flexing of the frame rails cracked the stainless steel, which required difficult repairs.

Solar Aircraft had produced just over 100 complete kits and probably three times that many in body parts when it reluctantly shut the operation down in late 1946. More Solars were built in the year after the war than Kurtis-Kraft midgets, but they failed to equal the Kurtis midget in competition. Most were Ford-equipped and therefore ran in the less expensive race classes, such as the United Racing Association Red Circuit in California and the American Racing Association in the East, both Ford clubs.

Elsewhere in the country, other builders were turning out midgets, but most were still making rail frame cars. Hiram Hillegass had been building midgets in Syracuse, New York, before the war using steering gears salvaged from the bankrupt Franklin Automobile Company. By 1945 he had set up shop in Allentown, Pennsylvania, with Roy Drumheller, another Pennsylvania Dutchman, and they were grinding out a simple chassis for the Ford engine. Floyd Dreyer was still building a few midgets in Indianapolis, as were the Reichenbach Brothers in Chicago; Louie Turco in Omaha, Nebraska; Fred Tomshe in Milwaukee; Cockeyed Brown in Kansas City; Felix Graves in Tulsa; Joe Silnes in Indianapolis; Ed Stone in Boston; and Johnny Wohlfiel in Pontiac, Michigan.

Buddy Chase in the Joe Brugnolotti Kurtis Offy, 1949. The author watched Chase take down a light from the top of a pole in the first turn at Morristown, New Jersey, in this car in 1950, with only a broken finger to show for the wild ride. After Brugnolotti sold the car it went to Ed and Bill Mataka, whose young driver, Mario Andretti, drove it to three feature wins in August 1963. *Bruce Craig*

Danny "Poison" Oakes in Kurtis midget serial number 4, delivered in early 1946. Note the rear radius rod design, unique to the earliest cars produced.

Dutch Schaeffer in the *Golden Arrow* Number 6 leads four Kurtis-Kraft midgets in action on the half-mile track at Dover, New Jersey, in October 1949. *Author*

Henry Van Rompay at Deer Park, New York, in 1947 in his Solar midget. This car was powered by a Ford V8-60 engine with a "Smith Jiggler" overhead exhaust valve conversion. The Smith heads put the exhaust valve overhead, thus getting the exhaust gasses out of the engine more directly and reducing the Ford's tendency to overheat, but as the intake valves remained in the block, there was little power increase. *Author's collection*

Harold Bicklehaupt at Hinchcliff Stadium, Paterson, New Jersey, in the Fred Maier Number 9, a Solar midget chassis carrying two two-cylinder BMW motorcycle engines hooked together. This car was quite successful in the East in 1948 and 1949, beating the Offys at Freeport, New York, three times in 1949 with Russ Klar at the wheel. The four-cylinder horizontally opposed air-cooled BMW engine anticipated the Volkswagen engine design that came into midget racing in the 1970s. *Author's collection*

Others were building midgets, but at nowhere near the pace of Kurtis-Kraft. Later Hillegass copied the upright Kurtis midget and produced a dozen or so tube frame cars, distinguishable chiefly by the integral cowl flare of their hoods.

The Reichenbachs built a copy of the Kurtis, distinguished only by the location of both front and rear radius rod connections opposite the U-joint in the driveshaft. By 1949 most of the other constructors were simply copying the Kurtis-Kraft.

In Glendale, Kurtis could not turn out cars fast enough. From the original batch of 10, Kurtis went on to build 44 complete cars in 1946 and kits or parts for 21 more for the buyers to finish themselves. By midyear he was making more than a car a week, in a business where 5 or 6 cars a year had been a lot of work before the war. Owners of prewar rail frame midgets and sprint cars bought Kurtis body parts with which to spruce up the looks, if not the capability, of their cars.

The rich purses at the Los Angeles Coliseum and at Hollywood's Gilmore Stadium attracted owners and drivers from as far away as the East Coast, and though the tracks already had large crowds, bringing in big names had its promotional value. For a 250-lap race in October 1946, White could advertise the presence of Duke Nalon, Bill Schindler, Johnny Tolan, Tony Bettenhausen, Myron Fohr, Buster Warke, Emil Andres, Joe Sostillio, Duane Carter, and Henry Banks, in addition to the local Southern California stars.

At the Coliseum 250-lapper, Schindler drove the black Number 2 rail frame car of Mike Caruso, his long-time car owner. Caruso, from Hicksville, on New York's Long Island, had been running midgets since 1934 and was one of the savviest of eastern car owners. Schindler was more than competitive in the Caruso car—he set fast time, started from the pole, and led the race until a six-year-old right rear tire blew. (Good rubber was still in short supply.) But he and Caruso failed to realize that the smooth-riding Kurtis midgets were about to make their best rail-frame cars outmoded. Caruso even had a new rail-frame Offy built for the 1947 season, but on the first day out it went up against three Kurtis Offys and was beaten, badly. Caruso quickly bought a used Kurtis-Kraft from mechanic Harry Stephens, and a new one from Kurtis-Kraft with a nice double-deck Burbank trailer to haul them. (Burbank Trailers were built by the nearby Burbank Trailer Company.) Stephens had not been successful with his car, but Mike Caruso did better. The

Johnny Carpenter at Hinchcliffe Stadium in 1949 in a Hillegass midget, a copy of the Kurtis-Kraft. Hiram Hillegass and Roy Drumheller built probably two dozen copies of the Kurtis midget at their Allentown, Pennsylvania, shop. They may be distinguished from Kurtis cars by the different configuration of the cowl; the slightly different radiator grille, oval with points at top and bottom; the steel "flag" on the rear bumper; and a few other details. Like many later Hillegass cars, Carpenter's used axles and hubs manufactured by the Progressive Machine Company in Allentown. This car, with the exhaust on the left, was probably powered by a Lea-Francis engine. *Bruce Craig*

Caruso *Deuce* and *Trey* would become the scourge of eastern midget racing for the next four years.

In California Nalon drove the Bill Krech Kurtis Offy, and Bettenhausen wheeled another Kurtis. Both of them realized the superiority of the design and spread the word back east. Bettenhausen told George Tuffanelli of Blue Lake, Illinois, who soon bought a Kurtis for Tony to drive.

The first Kurtis Offys known to have run on the Atlantic Coast included a pretty red and black Number 12 "Lion Head" car bought by Ed Bourgnon for Ted Tappet (Tappet's real name was Phil Walters, and under that name he went on to star in sports cars, driving for Briggs Cunningham, and to race at Le Mans, in France); the former Loren Bennett car, then owned by Frank Curtis of Hewlett, New York, another experienced eastern owner; and a third Kurtis-Kraft car Curtis assembled himself. All three showed up at Deer Park, on Long Island, on March 23, 1947. Chet Gibbons in the ex-Bennett car won his heat, the semi, and the feature.

In Harry Miller's day, 20 years earlier, one plant built the engine, frame, front and rear axles, steering gear, and many other parts. Bodies might call for a specialist such as Myron Stevens, but most cars were hand-built by a single builder, often with help from the buyer. After the war, as Kurtis specialized in frames and bodies, he developed dies from which to stamp body parts and jigs on which to build frames, but he was not equipped to do the foundry and specialized machine work to manufacture the steering, hubs, quick-change rear axles nor, of course, the engines that went into his cars.

Many of the driveline parts for the midgets had been designed by Leo Goossen for Offenhauser Engineering in the late 1930s and produced by Offenhauser's machine shop in small numbers. The production of racing

Jack Jiarusso in the Demuel Offy in 1941. This was one of Fred Tomshe's early attempts at using tubing to build a frame for a midget. Although Tomshe anticipated Kurtis' 1945 design, his were not particularly successful. *Author's collection*

Duke Nalon in Babe Tuffanelli's Number 22 Kurtis Offy in 1948. Tuffanelli is the heavy-set man with his hand in the pocket of his light-colored jacket, and his wife is wearing a fur coat. This was one of the earliest Kurtis midgets to reach the Midwest. *Jim Summers collection*

cars like the postwar midgets, where the numbers were high enough to achieve some degree of standardization, led to the creation of an industry of support manufacturers.

After the war Ted Halibrand started the large-scale manufacture of midget quick-change rear axles, using Model A Ford ring and pinion gears and change gear sets with ratios from 1.80:1 to 8.65:1. In the East, the Progressive Machine Company of Allentown, Pennsylvania, designed its own quick-change, using the Model A ring and pinion and its own pattern change gears. Halibrand and Progressive also made complete axles and hubs, with different mounting tapers. Casale, Schroeder, Evans, and Norden made midget steering gears, replacing the scrap yard Franklin passenger car units that had been used in racing cars for many years. Schroeder turned out brakes, torsion bars, stops, and steering arms. Bell Auto Parts provided air pumps, steering wheels, their own brand of midget brakes, and many other parts. (The standard midget wheel pattern was the six-bolt arrangement used on the Ford Model T.)

The Kurtis-Kraft midgets were designed for one of three engines: the powerful 102-cubic-inch Offenhauser, then selling for about $2,000; a hopped-up Ford V8-60 flathead engine; or the Drake, a modified, water-cooled version of the Harley-Davidson motorcycle engine. Fred Offenhauser had built 101 midget engines in the eight

Ted Tappet in the Eddie Bourgnon Kurtis-Kraft Offy at Deer Park, New York, August 27, 1946. Note the quarter-elliptic springs. This was also one of the rare midgets to have a tachometer. The car started out as Gib Lilly's Number 36 Ford-powered car, one of the first Kurtis-Krafts to be built. Bourgnon replaced the Ford with an Offenhauser engine and brought it east, where it was an instant success. Later torsion bar Kurtis cars outmoded the quarter-elliptic setup, and Bourgnon sold the car to Ken Hickey, who converted it to a transverse leaf spring front end. George Fonder drove the updated car with much success. Later Stan Frankenfield bought the car, painted it red and white, and gave it Number 2. It is the car that Tony Bonadies was killed in at Williams Grove when the rear axle broke. *Author's collection*

Tony Bonadies at the Freeport Municipal Stadium in 1948 in the Eddie Bourgnon Kurtis-Kraft Offy Number 5. Bonadies began driving midgets at Castle Hill, New York, in 1940 and became a star during the postwar midget boom. Tony placed fourth in AAA midgets in 1954 in the car the author now owns. He graduated to Championship cars but never qualified at Indianapolis. He was killed in a midget in 1964. *Author's collection*

Dee Toran in his spectacular blue and gold Kurtis Offy with the 7-11 dice on the tail, at the Thompson, Connecticut, Speedway in 1948. Toran drove a four-wheel-drive midget before World War II and won so many races that it was banned. Toran was jailed, briefly, for a crash in 1947 that killed driver Jeep Colkitt. *Author's collection*

years before Pearl Harbor; Fred and his successors, Meyer & Drake, turned out 51 in 1946, almost all for installation in Kurtis chassis.

By the fall of 1946, Kurtis still had an order backlog of more than 60 midgets. By the end of the year, he was turning out a rolling chassis every day. Photos taken of the shop make it look like an assembly line operation, though in truth it never achieved that level of standardization.

A buyer could specify the wheelbase, from 66 inches to 72 inches, with most built at 70 inches. He could specify the engine mounts, even if he was going to install the engine himself. He could have a pretty, formed Plexiglas windshield or a flared cowl, attached to a formed groove at the rear of the hood—another Kurtis-Kraft trademark. There was a choice of steering gears, rear axles, wheels, instruments, and suspension. Four–parallel bar suspension proved excellent on pavement, but on dirt, the choice often was to use a cross-spring in front. It was easy to swap the entire front axle and spring assembly between tracks if an owner chose, though most used torsion bars in the rear. A few cars had cross-springs front and rear.

Offy-powered cars came with a small grille opening, about 13 by 9 inches, while the harder-to-cool Fords got a larger opening. The same grille was used on the Kurtis 2000 Championship cars. On the midgets, Kurtis would often weld the buyer's initial into the front bumper.

A frame was built on the desired length jig and would be stamped with its serial number, as would the accessories, axles, etc. It would be put on a dolly and rolled to the next station, where the driveline and other parts were attached and the body panels hand-fitted. The bodies, while similar, were not interchangeable. It is rare for a body part from one Kurtis midget, even of the same length, to fit on another frame.

With an Offenhauser engine, a 12-gallon fuel tank, a 3-gallon dry-sump oil tank under the left side of the seat, painted and chromed, a ready-to-race Kurtis cost $5,000. The Burbank trailer was $500 more.

There were a few design changes after the first batch of cars went out the door. Partly for esthetic reasons and partly to give more driver knee room—the critical cockpit dimension—Kurtis raised the rear of the hood by 2 inches. There were also changes in the firewall at the front of the cockpit to improve room for the driver.

Production rolled on in 1947 at a rate just short of a car a day, not counting loose body panels and other parts sold for repairs or for the builders of "Kurtis copies." A total of 173 numbered rolling cars were sold that year and an estimated 76 kits, or cars in varying degrees of completion. Over at Meyer & Drake on West Gage Avenue in Los Angeles, 107 new midget Offenhauser engines were built, almost all of them for Kurtis chassis.

As an owner, Kurtis raced one of his own midgets in 1948 very briefly. Joe Garson was his driver, and the first time out, he tangled with Norm Holtkamp and landed in the grandstand. The car was repaired, but a few weeks later Garson and Holtkamp collided again. Garson flew 30 feet in the air, hitting a light pole. Garson was not injured, but Kurtis told the author, "I couldn't take it any more, and I sold the car to Bill Murphy, a Culver City Buick dealer, in 1949."

Offenhauser-powered cars were so superior to the Fords that even before the war there had been conflict. There were all-Ford clubs and the URA was divided into the Blue Circuit for overhead valve engines like the Offy and a Red Circuit for everyone else. In the East the A.R.A. was a Ford club, but gradually a few Offys were permitted in to satisfy the promoters. The advent of the Kurtis chassis aggravated the situation, because it allowed the Offenhauser engine to make full use of its superior power.

By the end of the year almost all of the Offenhauser midget engines had been pulled out of the old rail-frame cars and installed in Kurtis chassis. There were a few competitive Kurtis Fords such as Eddie Meyer's car that Gib Lilly drove and the Vic Edelbrock car that Rodger Ward piloted to a Gilmore victory on July 29, 1948.

Soon the similarity of the top cars became a problem. To avoid an all–Kurtis-Offy parade, Jake Kedenburg, promoter at Freeport, New York, decreed that no more than six Offys could run at those tracks on any one night. The prime Eastern Offy organization, the American Racing Drivers Club, was on the outs with Kedenburg. Though A.R.D.C. drivers ran at his tracks—they paid good purses—the A.R.A. was Jake's chief sanctioning group in 1946 and 1947. Still, the better Offy drivers threatened to make a runaway of Kedenburg's events. At one 1947 show, Kedenburg handicapped Henry Renard in an Offy half a lap. Not only did Renard start at the end of the inverted field, he had to start in the middle of the backstretch.

With the midget craze in full fury, men with more money than experience began moving into competition. Long Island's Frank Curtis catered to such customers.

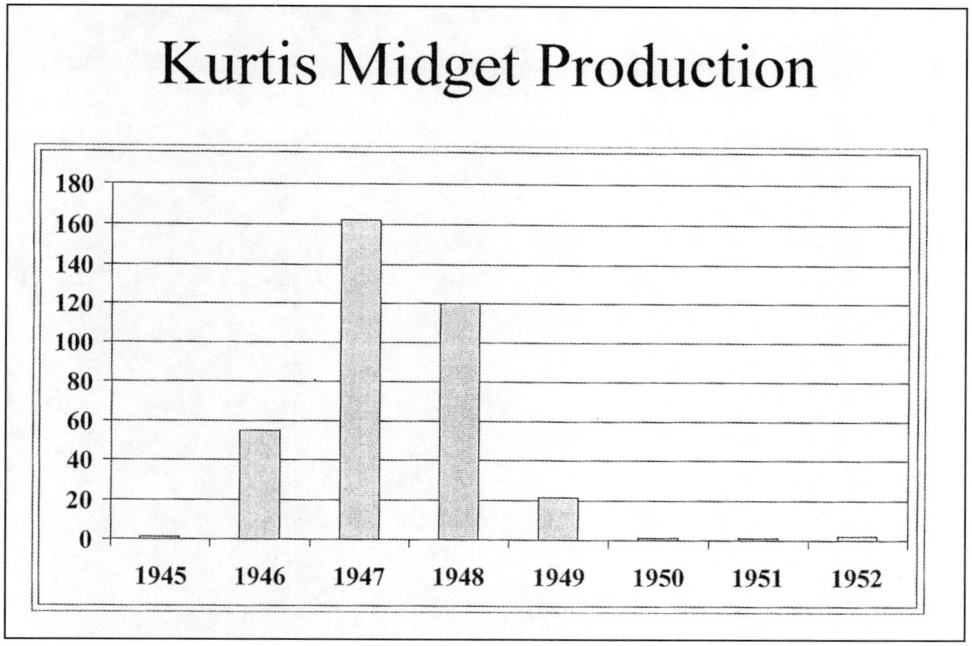

Kurtis midget production, 1945–1952. Kurtis built 55 cars in 1946, fewer than were produced that year by the Solar Aircraft Corporation, but as the superiority of the Kurtis design became apparent, buyers flocked to his Los Angeles plant. By early 1947, he was turning out almost a car a day. For the year he built 162 midgets and 120 in 1948. By 1948 the bloom was off the midgets and sales fell precipitously, almost vanishing in 1950. The collapse of the midget market came about not because there were competing car builders, but because fans quit going to see the little cars race in the huge numbers of the immediate postwar years. *Angela T. White*

Bob Sall tightens a spark plug on the Offenhauser engine in Johnny Ritter's Kurtis-Kraft midget in 1947, while Frankie DelRoy looks on. The four-cylinder Offy was the best midget engine ever built, at its displacement. Ritter was one of the most talented eastern drivers until he was killed in a bizarre infield accident at Medford, Massachusetts, October 12, 1948, while he was working on his car. *Author's collection*

Steve McGrath in the Frank Curtis Kurtis Offy Number 8 in 1948. Curtis, of Hewlett, Long Island, was one of the most active owners of Kurtis midgets. *Bruce Craig*

A crash at Denver's Lakeside Speedway in 1947, as half a dozen Kurtis midgets get together. Bud Shay is sideways in the Bus Bussard Offy Number 15, while Bert McNeese climbs Shay's hood in the Morris Musick Offy. Foster Campbell is right behind in the Ellis Evans Offy with Joe Giba in the Number 62. *LeRoy Byers*

He would buy a car on the West Coast, tow it east—racing and winning at tracks across the country—then sell it in the East as a proven winner for twice what he paid for it. Curtis bought and sold a dozen cars that way in two years.

The top midget owners and mechanics began to field teams of two or more cars. Jimmy James, a Los Angeles Dodge dealer, had a pair of Kurtis Offys, Number 28 and Number 48, with drivers such as Johnnie Parsons and Ernie Gessel. Buck Wheeler, a New York state tire dealer, had a team of cars, Number 21 and Number 62, driven by several drivers; the Macleod team had Bill Randall in Number 21 and Johnny Thomson in Number 20. Ed Stone had two Kurtis Offys, Jerry Willets a pair, and Frank Curtis two. Johnny Pawl, in Crown Point, Indiana, a former car builder himself, bought two Kurtises, Number 18 and Number 19. George Bignotti, a San Francisco florist later to become a noted Indianapolis mechanic, fielded a three-car team of Kurtis Fords, Number

Bobby Unser won three Night Before Pikes Peak races at Colorado Springs. He is shown with promoter J. C. Agajanian with his Doug Caruthers Kurtis Offenhauser midget. Bill Cantrell was its usual driver. *LeRoy Byers*

Action at the Englewood Speedway, Colorado. Eddie Jackson in Number 5 Kurtis Offy, Gene Pastor in the Chevy II Number 96, a Kurtis copy, and Speed Roberts in the Number 96 Kurtis Offy. The Number 5 here sports a fiberglass nose made by Charlie Peck to replace the original aluminum piece that was crunched in a crash. *LeRoy Byers*

47, Number 104, and Number 108. Ronnie Householder had a four-car team, three of them Kurtises. John Balch sold his first two Kurtis midgets to Howard Keck and had Curly Wetteroth build another pair for Danny Oakes and Andy Linden.

Dealers for Kurtis included not only Pawl, but Ed Walsh in St. Louis; Robert Wilke in Milwaukee; Frankie Del Roy in Paterson, New Jersey; Frank Spiller in Cleveland; and a few others. They took delivery of new chassis and in some cases finished them with paint, chrome, and upholstery, and installed and tuned engines.

There was racing six nights a week and on Saturday and Sunday afternoons across the country. Circuits existed in California, Washington state, Denver, Texas, Oklahoma, Missouri, the Chicago area, Indianapolis,

Lloyd Axel in his Kurtis midget in a match race against a track roadster. Despite the roadster's much larger engine, Axel won the match race with ease.
LeRoy Byers

Detroit, New York, New Jersey, New England, and even around Norfolk and Richmond, Virginia. It was truly a golden age of midget racing.

As an example of the Kurtis-Offy domination, Ted Tappet, driving different Kurtis midgets, won 13 features at Freeport and the track championship. The great drivers—Tappet, Schindler, George Rice, and their like—always stood out. Schindler won 53 feature races in both 1946 and in 1947, though Rice took the A.R.D.C. championship on points. But lesser drivers in Kurtis midgets found the playing field more even. The result was not better competition, but parades of Kurtis Offys, unable to pass one another.

In the East in February 1948, the A.R.A. club came to a "working agreement" with A.R.D.C., under which owners and drivers from both groups could run at each other's shows, but bickering between the Ford and Offy groups continued. At one race at Hinchcliffe Stadium in Paterson, New Jersey, promoter Ed Otto's prime track, the decline in crowds and purses led to a near strike in August. Ford owners objected to the number of Offys. Otto insisted on a minimum of 24 cars to put on three heat races, two semifinals, a consolation, and a feature. Only 18 cars, all Kurtis Offys, showed up and Otto canceled the show.

At the same time in the Southwest, a group of Offy owners demanded more money from promoters and struck when they were turned down. The shortage of good cars led to dwindling crowds and the closing of several midget tracks.

Offy midget Number 27, a 1946 Kurtis, originally serial Number 8, was bought by Lauren Bennett, who sold it to Frank Curtis, of Hewlett New York. This is one of the shortest Kurtis midgets built, with a 66-inch wheelbase. George Rice and Ted Tappet (whose real name was Phil Walters) drove the car for Curtis. American Racing Drivers Club President Dutch Schaeffer owned it for a time. Ray Oldenburgh, a collector, was killed in this car at a vintage meet in 1993.

Crowds began to fall off a bit in 1948, in part because there was too much racing. Too many tracks were running too close to each other. The fact that so many cars were virtually identical soured some of the fans. In the past there had been a rich mix of cars by different builders, powered by Offenhausers, Fords, Elto outboards, Carlheim cycles, Harley-Davidsons, Sowers, Jensens, Bugattis, Cragars, Millers, Sampsons, Universal Marines, J.A.P.s, Greens, Lea-Francis, Indians, and a profusion of other engines. Now it seemed most of the cars were virtually identical Offy-powered Kurtises.

When Pickles Bicklehaupt led the Kurtis Offys for a few laps in his ancient yellow motorcycle-engined Schloeder car it was more interesting than watching Ted Tappet, graceful as he was, dominate a race. Russ Klar's pretty Solar chassis Maier Cycle was an interesting alternative to the Kurtis domination.

Tappet drove one of the most exciting races ever seen at Freeport when his usual Offy-powered Kurtis failed to show up. Tappet jumped into a rail-frame car powered by a Universal Marine flathead four and simply out-drove a field of Kurtis midgets to win the feature. More such competition would have been good for the midgets.

By fall, the racing people realized they needed to stop squabbling and work together. At a benefit at Freeport for injured driver Johnny Ringger, the A.R.D.C. agreed to run alongside the A.R.A. cars. They put on two excellent races, one for Fords and the other for Offys. Art Cross won the Ford feature, Tony Bonadies took the Offy show. Nonetheless, the bloom was off midget racing.

Eastern driver Andy Furci in Ed Walker's Kurtis roadster midget at the 16th Street Speedway in Indianapolis in 1957. Furci was famous for breaking into "Back Home Again in Indiana" at the drop of a hat. *Author's collection*

Kurtis produced nearly 40 percent fewer midgets in 1948 than he had in 1947, a total of 108 complete cars and 45 kits. Meyer & Drake shipped 82 engines, 23 percent fewer than in 1947.

Overseas, in 1949, the Soviets blockaded the highways, canals, and railroads, providing the Western allies access to West Berlin, raising the threat of a new war. The world was changing. In the United States, television was becoming a commonplace appliance. No longer was it necessary to go out to the track for entertainment. You could find it for free and in comfort on the tube. Freeport and other tracks tried televising the races with Amalie Oil Company sponsorship, but camera techniques were in their infancy and the results were not satisfactory.

Many midget tracks simply closed, while some went to all stock car shows featuring junkyard Ford and Chevy coupes in demolition derbies. In 1949, sales of Kurtis midgets fell to 26 complete cars and 10 kits. Meyer & Drake produced only 19 Offenhauser midget engines. In 1950 Kurtis sold only a single car and Meyer & Drake 13 engines. The following year Kurtis again sold but one car, and only two Offy midget engines were built.

There were still orders for replacement parts, but those were more trouble than they were worth to fill, and Kurtis decided to drop the small car business entirely. He cast around for a buyer and in early 1953 found Johnny Pawl. Pawl had been a Kurtis dealer and had bought (with financing help from Murrell Belanger) Bob Wilke's (the owner of Leader Card Industries in Milwaukee) Offenhauser and Kurtis dealerships. Wilke and Belanger helped Pawl buy the Kurtis-Kraft midget manufacturing fixtures, parts, supplies, and two unfilled orders, at a price of $7,500 to be paid over two years.

Pawl produced the rest of the Kurtis midgets built, though very few complete cars went out the door of his Crown Point shop. Once the frames that came from Kurtis were exhausted, Pawl's business was chiefly in repair and service parts.

For Kurtis, the midget decline coincided with the building of the first Indianapolis car he designed himself, and his first serially produced Championship car, the KK2000, essentially a beefed-up version of the midget. An innovator, Kurtis was not terribly saddened by the end of the midget boom; it gave him a chance to concentrate on Indianapolis, the real prize in the racing world.

MIDGET ROADSTERS

Kurtis made one more stab at building midgets in 1956, when the little cars were having a resurgence of popularity. He had been building Indianapolis roadsters for four years and decided to apply the same ideas to the smaller cars. That fall, more out of nostalgia than financial interest, he designed a roadster version of a midget. Its Offenhauser engine was offset 5-1/2 inches to the left, and the driveshaft was run alongside the driver, allowing a lower center of gravity and better weight distribution for turning left. The Kurtis midget roadster stood only 26 inches off the pavement, but was 30 pounds heavier (900 pounds, dry weight) than a conventional upright Kurtis, and weight was important on a short track where every ounce counted. Designated the Kurtis M-100 series, 10 were begun.

A pair of Offenhauser-powered Kurtis-Kraft midgets in the East in 1956. Chuck Arnold is in Ken Brenn's Number 24, the former Johnny Ritter car, and Freddie Meeker is in Fred Kiernan's Number 17. *Author's collection*

The Kurtis midget roadsters were similar to the KK500G2 Indianapolis cars, carrying a hood scoop covered with pierced metal mat and a protective bar to keep the driver's right elbow off the rear tire. The nose scoop was a smaller version of that on the Indy car and the suspension and steering were also similar.

The first numbered midget roadster car was bought by Chicago's Ashley Wright in December 1956. (Ray Tomaseki actually took delivery of his car, Serial Number M-102, first. Shorty Templeman was its driver.) Bob Gregg, in the Wright car's first race, on the paved track at Saugus Speedway in California, set fast time, but wrecked in his heat race. Four others were sold to Ed Walker, Tassi Vatis, Ernie Alvarado, and Hank Green.

CAR SPOTTER'S GUIDE TO THE KURTIS MIDGETS

The upright Kurtis midgets were built to similar plans, though the wheelbase could vary by 6 inches and different engines could be installed. The feature that distinguished the Kurtis uprights from most of the copies was the groove at the rear of the hood where the windshield was attached, though George Shilala, shop foreman at Kurtis, built a number of identical midgets after he left Kurtis-Kraft.

Ten exceptions were the Kurtis roadster-style cars built in 1956 and 1957. They may be immediately spotted by the fact they are much lower and slightly wider, with offset engines.

The roadster midgets proved to be good on pavement but not on dirt. There was too little suspension travel to handle rough tracks, and as soon as the front axle bottomed on the frame, the car would go into severe understeer and become uncontrollable. The problem was correctable, but after several business reverses, Kurtis' interest in redesigning the roadster midgets disappeared.

The four unfinished chassis were sold, three to Charlie Peck and one to Ron Ward, a Kurtis employee. Peck had a small business building fiberglass replacement parts for Kurtis midgets including hoods, tails, and a shovel-shaped nose with a horizontal grille. Peck put his own fiberglass noses and tails on the three cars and resold them. All are believed to still exist.

Ashley Wright's midget roadster was quite successful, though only on pavement. It was driven to a number of wins by Len Sutton, Shorty Templeman, and Tony Bettenhausen. Sutton won a pair at the 16th Street Speedway in 1957 and 1958, including the last "Night Before the 500" in 1958. Bruce Homeyer bought the car at the end of 1958 and ran it as the *Konstant Hot Special*, and subsequently sold it to Tony Bonadies.

Bonadies' roadster stayed in his possession. His son, Tony Jr., crashed it in competition and at last report still had the remains.

Four Kurtis-Kraft Offy midgets race for position at Seekonk, Massachusetts on May 25, 1957. Johnny Kay in the Nanigan number 3 leads Freddie Orlando in the Paul Young number 41, Bobby Boone in Swede Anderson's number 21, and Jerry Russo in the Scrivani number 21. *Author's Collection*

Anastasis "Tassi" Vatis, who bought the third roadster, gave it the number *9* and painted it in the same red and black color scheme as his Kurtis upright, a car the author now owns. Bonadies was the usual driver for Tassi's roadster, and he almost always ran it on pavement. Tony was killed in the Frankenfield dirt car at Williams Grove in 1964, and the Vatis roadster was sold to Pop Sincerbaux and then to Freddie Charles Heydenreich.

WOULD THEY RUN FOREVER?

Kurtis midgets continued to compete long after Kurtis-Kraft itself was only a memory. Rodger Ward won the famous Formula Libre race at Lime Rock, Connecticut, against the world's top sports cars in 1959 with a 10-year-old Kurtis upright midget. The author set the U.S. midget speed record at Bonneville in 1989, in a 41-year-old Kurtis.

Kurtis chassis made up the majority of the competitive midgets well into the 1970s and many would have lasted much longer, had it not been for changes in engine rules allowing larger stock-block engines. Fitting a Ford Falcon six into a Kurtis chassis required major butchery. The horizontally opposed Volkswagen engines used in the early 1980s were not a good fit in the standard Kurtis frame.

George Shilala would continue on to build perhaps 50 cars to the Kurtis pattern after he left Kurtis-Kraft and Don Edmunds built 200 cars, though all but a handful were only kits. Hiram Hillegass was more successful with his dirt sprint cars than with his midgets. He built several sprinters for the central Pennsylvania gang that ran at Williams Grove, Reading, and Allentown. Most notable of these was Tommy Hinnershitz, the farmer who was a many-time champion on the dirt.

As time went on in midget racing, the Kurtis chassis were updated with roll cages and kits to install crossed torsion bars behind the rear axle to extend the spring base, as on the Indianapolis cars. Elsewhere a great many copies of the Kurtis were built, and in fact virtually all tube frame midgets built after World War II were close copies of the Kurtis, many using Kurtis body parts or the Charlie Peck fiberglass noses and tails.

For more than a quarter-century most midget racing had been done in cars of Kurtis design. Somewhere today, there is probably a Kurtis midget still struggling to compete, half a century after it rolled out of the shop in Glendale.

Tony Bonadies in the Pop Sincerbaux Kurtis roadster midget, the ex-Tassi Vatis number 9 car. Tony had driven this car and an upright Kurtis number 9 midget for Vatis also. The roadster is now in the collection of Freddie Charles Heydenreich. *Author's Collection*

Kurtis Champ Cars

CHAPTER 4

During World War II Kurtis' ideas for Indianapolis cars were put aside while he did defense work, though he continued to daydream over postwar plans. In 1945 his wildly successful new midget occupied most of his time, though Championship car owners came to him for chassis repairs and with orders to build new cars to their own specifications.

At Indianapolis the Speedway was still a weed-choked ruin. But in September 1945 Wilbur Shaw interested Indiana businessman and baking soda magnate Anton "Tony" Hulman in buying the track. In early 1946 Hulman moved into the Speedway and began its restoration. There would still be an Indianapolis 500 in the postwar world. Car owners began to prepare.

In Los Angeles, Ross Page had bought the former Leon Duray Barbasol car, and he asked Kurtis to build a new tube frame chassis, replacing Ernie Weill's rail frame handiwork. (A rail frame for a race car was built of channel-section steel, much like that of a standard passenger car. Kurtis did not invent, but he developed, much stiffer frames built out of aircraft-quality tubing.) Lewis Welch went to Kurtis that winter with plans for a dramatic new car to carry his powerful Novi V-8 engine.

In building the Ross Page car, his first tube frame Indianapolis car, Kurtis set the engine 2 inches to the left, allowing the driver to sit alongside the driveshaft and permitting a lower center of gravity and a lower profile. This was an extension of the idea Harry Miller had followed with his front-wheel-drive cars, 20 years earlier. Duesenberg built an offset-shaft car in 1927, but it started no trend. (The Duesenbergs used a pinion gear set at an odd angle to the ring gear, and it did not last 500 miles.) Indianapolis mechanics had noted however that when the riding mechanic was finally eliminated in 1938, the narrower one-man cars were taller than the two-man vehicles had been, because the driver had to sit over the shaft. Kurtis was trying the offset idea once more. Based on his wartime experience in building canopy forms for aircraft, Kurtis built the headrest for the Ross Page car out of plastic, a first for an American racing car.

At the start of the 1952 Indianapolis 500, Jack McGrath in the Hinkle number 4, a KK3000, jumped out to a lead over Jim Rathmann in the Grancor number 59, another KK3000. Duane Carter in a Lesovsky chassis car followed, as Johnnie Parsons in the 1950-winning Kurtis and Cliff Griffith in the Sarafoff KK2000 trail in fourth and fifth positions. Freddie Agabashian, the pole sitter, in the Cummins Diesel, fell to 8th because the diesel was unable to accelerate as rapidly as the Offenhauser-powered cars. *Author's Collection*

Mel Hansen in the Ross Page Number 41 in which he finished 11th at Indianapolis in 1946. This was the first offset-driveline car Kurtis would build, and as historian Mark L. Dees said, "It was a proto roadster." *Tom Saal collection*

The use of tubing for the frame was not entirely new. Fred Tomshe had built a midget out of tubing for the Marchese brothers in 1938, but he simply treated the tubes as frame rails. The Tomshe three-spring suspension was obsolete even for 1938, and the design had nowhere. Kurtis' car, however, was the first step in a near revolution of chassis design. Historian Mark L. Dees called the Ross Page car a "proto roadster."

Kurtis wanted to tilt the engine in the Ross Page car to the right by 45 degrees, as Harry Miller had done in 1938 with his Gulf-Miller cars, to further lower the car's profile. Conventional wisdom at the time was that the idea would not work, and Page refused to allow Kurtis to build the car that way.

That same winter and spring, the Kurtis shop also completed rebuilding Ed Walsh's wrecked 1941 Indianapolis car. George Connor qualified it 30th at Indianapolis in 1946 and finished 24th with a burned piston. The car ran at Langhorne, Goshen, Milwaukee, and then Atlanta, where it won the dust-plagued 100-miler in which George Barringer and George Robson were killed.

The Novis

The great and powerful, but star-crossed and tragic, Novi racing cars had come about in 1940, when Lewis W. Welch, a parts subcontractor to the Ford Motor Company and a friend of Henry Ford, went to W. C. "Bud" Winfield and asked him to help design a V-8 engine with which to conquer Indianapolis. Winfield had helped design supercharged engines for Robert Bowes, and Welch had Ford connections that many believed included surreptitious financing of his racing efforts with V-8 engines.

Harry Miller's former great designer, Leo Goossen, was brought in to do the engineering and drawings for a supercharged, twin-cam V-8 engine. Fred Offenhauser's shop had the engine's castings done and machined. Welch, Winfield, Marvin Jenkins, Pete Clark, and Tony Morosco assembled the engine, and Offenhauser installed it in a Miller-Ford chassis that had run at Indianapolis in 1935.

The engine was a sensation at Indianapolis in 1941. The sound of its supercharged V-8 shook the old wooden stands to their foundations. It quickly became obvious that the six-year-old frame was not strong enough to let out the full 500 horsepower of the supercharged engine, and legend has it that a block was put under the gas pedal. As a result, it was not the fastest car on the track; in fact only three cars qualified slower than its 120.653 miles per hour, but Ralph Hepburn drove it to fourth place. World War II intervened before the car could be developed further.

As the war ground to a close, Welch, Goossen, and Winfield planned a new car to carry the engine. In 1945 front-wheel drive seemed the way to go, and Lou Moore was about to prove it with three victories in a row—1947, 1948, and 1949—with his Blue Crown cars.

The chassis Goossen drew up had a fairly conventional rail frame, with independent front suspension, using longitudinal torsion bars with a solid axle and leaf springs at the rear. Wilbur Shaw's 8CTF Maserati had used torsion bar suspension, winning the 1939 and 1940 races, but while Welch turned to Frank Kurtis to build the car, he refused Kurtis' suggestion to build it with a lighter, stiffer tubing frame with torsion bars all around.

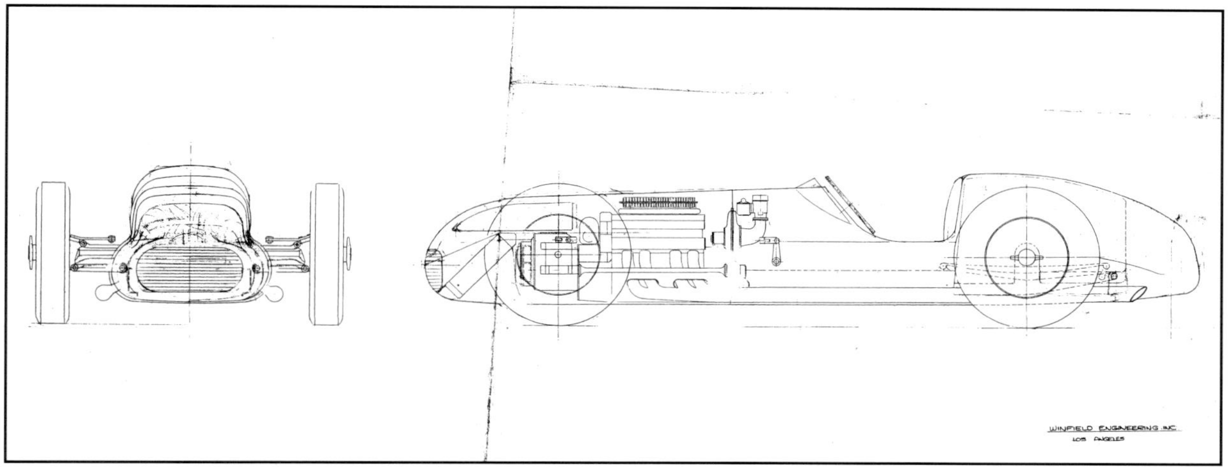

Leo Goossen's drawing of the Novi. *Leo Goossen Archive*

The first Kurtis Novi as it rolled out of the shop at the end of April 1946. Tony Morosco is in the driver's seat, while Frank Kurtis, Marvin Jenkins, and owner Lew Welch stand behind the car. *Marvin Jenkins collection*

EDWARD J. WALSH JR.

Edward J. Walsh Jr., a millionaire investor, was an enthusiastic, talented racing driver. He started driving race cars in 1927, and he later won a season championship in a midget. If he had continued a professional racing career after 1937 he might have won the Indianapolis 500 as a driver, rather than as a half-owner of Johnnie Parsons' 1950 winning Kurtis-Kraft.

Walsh was born in St. Louis on February 2, 1909. A descendant of Rene Chouteau, one of the founders of the city in 1764, and of Edward Walsh, who came to America from Ireland in 1818, he was the son of wealthy parents. Walsh said in later years that the family fortune was founded on selling liquor to the Indians. His father, Edward Walsh Sr., was an engineer who had worked with the Dorris Motor Company in St. Louis. The senior Walsh was also an influential Catholic layman. His family gave the land on Oakland Avenue to St. Louis University, a Jesuit college, to build a field for its athletic activities. Those sports came in the 1930s to include midget auto racing. Ed Jr. graduated from the university in 1931.

Auto racing attracted Walsh early. He first went to Indianapolis in May 1927 at the age of 18. At home he began driving a Model T Ford speedster at Don Zeiter's Creve Coeur Lake Speedway in St. Louis, a half-mile dirt oval that closed in 1935 when the grandstands blew down in a tornado. In 1935, at the age of 26, he tried his hand at the then-new sport of midget racing.

"I was a professional in those days," Walsh told the *St. Louis Post-Dispatch* in 1971," but in those years there wasn't much money to be won. In the 1930s, a pro could clean house at a place like Walsh Stadium and only win $90." He drove an Elto 4-60 outboard-powered Dreyer to the Missouri State Championship in 1936 and then the Walsh Stadium Championship. He raced indoors in the winter at the St. Louis Arena. Racing the small cars was immensely popular in St. Louis at that time. Dizzy Dean and Pepper Martin of the Cardinals baseball team owned midgets, and Dean even drove his until the team put a stop to it. Walsh, Dean, and Martin would sit around with Lee Hammock, another local racer, in Hammock's home in Brentwood, talking racing.

Since Walsh had the money for both fast cars and fast boats, he also bought and piloted one of the galloping shingles that were outboard racers in the late 1930s, and he enjoyed his own yacht on the Mississippi. Some of his friends began calling him "Steamboat Ed."

Walsh retired from driving midgets at the end of the 1937 season, when he got married. The midgets had become too dangerous by then. Roll bars and other safety devices were yet to be invented, and Walsh decided that for a married man it was a bit too risky.

In 1938 he went to Indianapolis as a fledgling car owner and leased a car from Joel Thorne. It was a 1932 Stevens chassis with a 255-cubic-inch Offenhauser engine that Bill Cummings had driven at the Speedway six years earlier. Labeled the *P R & W Special* Number 58, the car qualified 30th at 116.339 miles per hour, but Billy DeVore brought it up to 8th place by the end of the 500 miles.

The ex-Schrader car, sold in 1933 after Miller's bankruptcy, ran in at least two European Grands Prix in 1934, then was parted out, with the engine going into Bunny Phillips' Type 35 Bugatti and the running gear into Phillips' junk pile. Lou Moore bought the frame and converted it to a rear-wheel drive, with a 255 Miller engine and the cross-spring independent suspension setup seen here. Ed Walsh bought the car from Jack Holly in 1939. When Walsh hired Kurtis to repair it, it began an association that was to continue through the rest of Frank's career as a car builder. *Mark L. Dees collection*

Such success his first time out emboldened Walsh to return to Indianapolis in 1939 with a Miller chassis, built by Stevens, the ex–Four Wheel Drive car that Gus Schrader had wrecked on the seventh lap in 1932.

This was a famous car, whether or not Walsh appreciated the fact. It had sat, unraced, at Harry Miller's Los Angeles plant in 1933 while Miller tried to stave off bankruptcy. When Harry's business finally came to its end that summer, Frank and Al Scully bought it at auction for $1,600. The Scully brothers hired Peter DePaolo to drive it in several European races in 1934, including the Tripoli Grand Prix, where it finished in seventh place before an audience of Italian dignitaries that included Benito Mussolini. DePaolo drove the local Italian governor, Italo Balbo, around the course in the car. They then took it to Berlin for the Avusrennen race. There the Miller engine threw two connecting rods through the hood, one landing not far from Adolf Hitler, who was a spectator.

The Scullys returned the car to Los Angeles, where it was taken apart. The engine and some other parts, probably including most of the four-wheel-drive system, were sold to Bunny Phillips, who, in 1939, put the V-8 into a Bugatti type 35 chassis. That car finished 13th at Indianapolis in 1941 and still exists.

The FWD chassis was sold to Lou Moore, who, with Harlan Fengler, put a conventional rear-drive setup in it with an odd independent front suspension. Fred Frame finished 32nd with it in 1936, when its 255-cubic-inch Miller engine broke a piston. Jack Holly bought the car in 1938 and installed an Offy engine for Al Miller. As the DoMont's Pepsi-Cola Special Number 55, it finished 18th at Indianapolis.

The Indianapolis crowd must have thought they had a real pigeon when Walsh showed up in 1939 looking for a race car. He bought the Jack Holly Miller and ran it that year as the *WBW Special* Number 56. He had a good driver in Floyd Davis, who would win the 500 two years later. Davis was able to qualify it 29th, and finished 27th when a shock absorber on the old car failed on the rough Indianapolis bricks.

Walsh met a young Frank Kurtis in Indianapolis that May, 1939 when Kurtis was called into service to repair Billy DeVore's wrecked *Barbasol Special*. Impressed, Walsh asked Kurtis to go to St. Louis to see what he could do about improving his car. Kurtis rebuilt the old Miller and worked for Walsh's crew at the Speedway in 1940 when the car, then the *Falstaff Special* Number 26, driven by Louis Tomei, finished 16th after the exhaust pipe fell off.

After the 1940 race, Walsh hired Kurtis to build a new car for 1941, using the engine and what parts he could off the old one. The car was the *Tom Joyce 7-Up Special* Number 26, the first Kurtis to race at Indianapolis. Sam Hanks, who drove the car in 1941 at the Speedway, qualified 32nd but crashed on Carburation Day when the crankshaft broke. The wreck went back to Los Angeles with Kurtis, but before it could be repaired World War II descended upon the United States. The racing shops in California turned to defense work, and Walsh joined the Navy, serving in the Pacific and emerging in 1945 as a commander.

In 1946 the repaired car returned to the Speedway and driver George Connor took it 38 laps until a piston broke. A year later, in 1947, the car went 32 laps, until the fuel tank began to leak. Walsh then sold the car to Mutt Anderson and it passed out of the Kurtis story.

Walsh, visiting California to see to the repairs to the Indianapolis car at a time when Kurtis' midget business was booming, suggested that he and Frank become partners. Walsh would put up half the capital—$35,000—and become half-owner, while Kurtis would run the shop and draw a salary. They would split any profits.

The Walsh partnership was not a good move for Kurtis. He was doing very well producing midgets and did not need Walsh's investment. To Ed Walsh, racing was a plaything and perhaps a way to finance his hobby. Walsh was used to traveling first class and enjoying only the best. Kurtis was the son of a blacksmith who had a hardscrabble upbringing and no education in what Walsh felt were the finer things. At Indianapolis in May, Walsh would live in a suite of first class hotel rooms while Kurtis put up in a boardinghouse in the middle class village of Speedway. Walsh would eat well and drink well, at the company's expense. It was a tax writeoff to him, but to Kurtis it was his livelihood.

Lou Moore converted the ex-Schrader four-wheel-drive car to rear-wheel drive and replaced the front drive axle with this cross-spring independent suspension setup. It cannot be said to have worked very well, though the car qualified for the 500 in 1936, 1938, and, finally for Ed Walsh, in 1939. *IMS*

For a while Walsh and Kurtis worked together. Walsh was the gladhander, Kurtis the inside man, content to stay out of the limelight and build race cars. When Johnnie Parsons drove the *Kurtis-Kraft Special*, in 1949, finishing second at Indianapolis and winning the 1949 AAA National Championship, Walsh was ecstatic. When Parsons won the 500 in 1950 Walsh proclaimed himself the car owner, which, while half-true, rankled Kurtis.

After the 1950 race, Walsh bought Kurtis' own KK3000, the *Wynn's Friction Proofing Special* that Freddie Agabashian drove, finishing 28th with an oil leak. Ed entered a used KK4000 at Indianapolis in 1952 and in 1953 Sam Hanks won the National AAA Championship, driving the Walsh KK3000 as the *Bardahl Special*. Walsh bought another used KK4000 and a new KK500C in 1954. But by then he was no longer a hands-on car owner. The KK500C he entered at Indianapolis in 1954 as the *Bardahl Special* Number 38 was hardly a satisfying experience. Bill Holland, a driver in the twilight of his career, could not get the car up to speed and spun trying to qualify it. Jim Rathmann put the car in the show 28th, and in the race Pat Flaherty, driving relief, spun on lap 110 and wrecked it. Although Rodger Ward finished 8th in the car in 1955, Walsh, as he said later, was not allowed to touch it. After the 1955 race, he sold the 500C to Fred Sclavi and never again worked in the garages or the pits at Indianapolis.

EDWARD J. WALSH JR. (CONTINUED)

In 1952 Walsh began to race with the Sports Car Club of America. He bought a Porsche Carrera and drove in such events as the Nassau Speed Week at Windsor Field in the Bahamas. Walsh found SCCA racing at places like Mid-America Raceway in Wentzville, Missouri, more satisfying than the $150,000 a year it cost then to enter a car at Indianapolis. Indianapolis had become too expensive and too impersonal for his tastes. He became active, indeed a power, in SCCA.

Over the years, Walsh owned 20 race cars. Of Indianapolis he said, "In the beginning there was a lot of involvement. Sam Hanks and I used to do our own mechanical work, but then it evolved into an impersonal thing. In the later years," he told the *St. Louis Post-Dispatch*, "I never got to touch the car. I never took off my suit coat, never went into the pits. There wasn't much pleasure in that."

In 1955, the Walsh-Kurtis partnership was dissolved. Walsh said he felt Kurtis was not making enough profit on their investment. Ed Walsh was ever more involved in civic affairs and did most of his racing in Missouri. He became the vice president of the SCCA. In 1956 he and SCCA President Jim Kimberly devised the idea of sending sports racer Carroll Shelby to Indianapolis. They ordered a Kurtis model KK500G2 Indy roadster, bought a brand-new 255 Offenhauser engine, hired Harry Stephens as chief mechanic, and were ready to see their sports car legend tackle the Speedway, but then those plans fell apart. Shelby was busy racing in Europe. The new Kurtis sat, untested, in its Indianapolis garage. Eventually they sold the car to John Wills, who ran it later as the *Hoover Motor Express Special*. It later went to Roy McKay and then to Ray Brady, but it never qualified for the 500.

Walsh became SCCA president in 1958. He continued to race on the road courses but he never returned to Indianapolis as a car owner. He was usually at the Speedway each May to kick tires and talk to his old friends there, men like Jack Hinkle and Bob Estes. He campaigned several Kurtis sports cars, including an Olds V-8–powered Kurtis KK500S in October 1955, with which he beat two Ferraris and a Jaguar in the B-Modified race at Smartt Field in St. Louis. Ed had Kurtis build one of his last sports cars in 1962 to carry a noisy two-cycle, three-cylinder Saab engine. The independent-suspension car was a break with Kurtis' usual solid-axle designs. With it Walsh won the Class H-Modified SCCA Championship in 1964. In 1971 he owned a Bobsy Ford. Later, in the late 1970s, he owned a Suzuki-powered Ocelot sports racer in SCCA Class D. While he had drivers, he kept up his SCCA competition license. In 1976, when driver Jim Boehm was unable to drive at a Mid-America raceway event, Walsh, then 68, took the car out himself and won his class with speeds of more than 130 miles per hour.

Ed Walsh, mechanic Harry Stephens, and Frank Kurtis are seen here in 1950 with the two Wynns-sponsored Kurtis-Krafts, the Indy-winning car that Parsons drove on the left and on the right the KK3000 that Freddie Agabashian drove. *Carmen Schroeder collection*

At Indianapolis in 1946, Ralph Hepburn qualified the Novi at nearly 134 miles an hour, although the car went out of the race with engine trouble on Lap 121. As the builder, Frank Kurtis won applause for its construction, and it helped boost his reputation as a builder of Indianapolis cars.

After the 1946 race Ed Walsh persuaded Kurtis to let him invest in Kurtis-Kraft as a partner. Kurtis had been working on Walsh's Indianapolis cars for four seasons and the wealthy St. Louisan was willing to put money into the business. It seemed like a good idea at the time, but Kurtis would come to regret it.

Lew Welch asked Frank to build a second, virtually identical, Novi for 1947. The second Novi was one of only two Indy cars Kurtis-Kraft built for the 1947 Indianapolis race, as construction of nearly 250 complete midgets and midget kits was all the plant could handle. It is possible another Champ car was built off the books, and Kurtis said he repaired some crash-damaged cars. Kurtis employees Ed and Zeke Justice took on some of the repair jobs in their own shop after hours, and eventually Kurtis told them to stop dividing their time between Kurtis-Kraft and their night jobs. Ed and Zeke quit, and became very successful marketing their own brands of automotive chemicals.

At Indy in 1947, Cliff Bergere finished 21st in the old Novi and Herb Ardinger came in 4th in the new car.

Winfield, Welch, and Marvin Jenkins had tested both Novis at Muroc dry lake, 125 miles out in the desert from Winfield's home in La Canada. They took the cars up to Muroc on a flatbed truck, in the days when Edwards Air Force Base was a fraction of its present size, and ran them on the lake bed. On one trip the cars went by road, but Marvin Jenkins, an air transport pilot, flew Winfield and Welch up in his twin Beech. They had found a strip named Pancho Barnes Airport on the CAA sectional chart. Not knowing much about the field they were amazed when, after they landed, a Cadillac limousine from Barnes' Happy Bottom Riding Club came to pick them up.

Florence "Pancho" Barnes was an early aviatrix and a pal of Amelia Earhart, who ran a bar and party resort on the desert for fliers from Edwards and sporting types from all over the West. Usually men who flew in to her air strip were looking for more than just a place to test a car. Jenkins, an upright Mormon from Salt Lake City, where his father was mayor, had to tell Pancho that their intention was only to test the Novi.

In 1947, after the Indianapolis race, the Mobil Oil Company (Standard Oil Co. of New York) offered to sponsor straightaway speed record attempts at the Bonneville Salt Flats in western Utah. Kurtis built a special, streamlined cockpit canopy with a high, finned tail that was attached to the 1946 car with Dzus fasteners. Marvin drove the Novi to a 179.434–mile per hour International class D record, a bit of a disappointment to the team, which had hoped to exceed 200 miles per hour.

The two Kurtis-built Novis ran at Indianapolis for Welch through 1955 without finishing better than third (in 1948). Ralph Hepburn and Chet Miller died in identical accidents in the 1946 car in 1948 and 1953. Duke Nalon was seriously burned when his Novi broke its rear axle and crashed in 1950.

Kurtis repeatedly told Welch that the front-wheel-drive chassis were inadequate for the 500-horsepower Novi engines. In 1949 Kurtis took Welch up on the overpass and pointed out to him how the Novis' smoking front tires spun halfway down the backstretch because of weight transfer as the cars accelerated. In 1956 Welch

The 1946 Novi with its straightaway canopy at Bonneville in 1947, where Marvin Jenkins, son of Ab Jenkins, drove it to International Class D records. This was Kurtis' first experience with high-speed runs at Bonneville, and he built the cockpit canopy and fin to streamline the Novi.
Mark Dees collection

The remains of the 1946 Novi after Chet Miller's fatal crash at Indianapolis during practice in 1953. The same car killed Ralph Hepburn in 1948. It was retired after Miller's death and lay, unrestored, for 40 years. Both Hepburn and Miller got into spins, tried to recover, and ran nearly straight into Indy's wall.
Author's collection

Tony Bettenhausen in Kurtis-built *Novi Air Conditioner Special* Number 27, one of the KK500F models built for Andy Granatelli. Tony finished 15th. *Armin Krueger*

finally gave in and had Kurtis build two new tube frame rear-wheel-drive cars.

With Walsh's investment and the income from midget production, Kurtis had more money behind him than he had ever enjoyed before. Kurtis began to dream once more of building an Indianapolis car to his own ideas. In April 1947, he began drawings of a car he called the Kurtis-Kraft Model 1000. It would not take form until 1948, but would be the second most successful Championship car of the 110 his company would build over a 23-year span.

The 1941 Walsh Kurtis finished 26th at Indianapolis in 1947, its last year under Walsh's ownership. Mutt Anderson bought the car after Indy but was never able to get it up to a qualifying speed again. Buster Warke then bought it from Anderson and ran it in Eastern AAA competition. The car went on to the NASCAR Speedway Division in 1952–1953 and disappeared somewhere in the Midwest in the late 1950s.

THE BOWES INDIANAPOLIS CAR

In 1938 Lou Meyer had commissioned Myron Stevens to build a chassis for a supercharged straight-eight engine designed jointly by Bud Winfield and Leo Goossen, incorporating timing gears at the flywheel end of the engine driving a centrifugal supercharger at the front. The chassis was a beast to handle even for as tough a driver as Meyer, or later, Rex Mays. In 1948 Robert Bowes, the sponsor of the car (and its owner after Meyer retired), had Kurtis build a new chassis for the engine. His chief mechanic, Pete Clark, insisted on a traditional rail frame design with a cross-spring front suspension and short torsion bars bonded in rubber at the rear. Kurtis urged Bowes to let him build a stiff frame of tubing with torsion bars at both front and rear, turning in bushings at the wheel end and retaining them in an adjustable fitting at the other. Clark would not listen, and the result was little better than the 1938 Stevens car and far uglier, though some of the earlier body parts were used.

Even so, Mays finished sixth at Indy with the Bowes car in 1947 and put it on the pole in 1948. Afterward, Mays and Clark split. At Indy in 1949 Rex drove one of the Novis, and for the dirt track races took over Erwin Wolfe's Kurtis 2000, the car in which he was killed at Del Mar that fall. Franklin Merkler bought the Bowes car, but in five attempts it never again qualified for the 500.

NOVI AUTOMOBILE AIR CONDITIONER SPECIAL
FLYING A
27
BOB ESTES SPL

Jackie Holmes, in the Clancy six-wheeler Number 57, leads eventual winner Myron Fohr in the *Marchese Special* Number 2 at Milwaukee, June 2, 1949. Clancy's Kurtis was essentially a midget, stretched to Championship length. The six wheels worked pretty well, but the added weight kept the car from being a winner. These were the days when fences were wood, men raced in T-shirts, and Championship races were run mostly on dirt. *Armin Krueger*

PAT CLANCY SPECIAL

Pat Clancy of Nashville, Tennessee, had an idea that if there was more rubber on the ground the added traction would make a winning car. Clancy bought a stretched-out midget frame from Kurtis as well as stamped but incomplete body panels. He had his mechanic, A. J. Bowen, put together a three-axle, six-wheel car. A somewhat similar idea had been tried on a midget with dual rear wheels in 1938 by Detroit's Lloyd "Booby" Logan. The English Tyrell team, using the same reasoning, would later build a Formula One car with two steered front axles. The idea did have some merit, and the Clancy-Kurtis six-wheeler car qualified at Indianapolis twice, in 1948 and 1949. But despite using the first magnesium wheels, the weight of the extra axle hurt the Clancy car's competitiveness, and for the 1950 race it was rebuilt with the conventional two axles. It qualified that way and finished 17th. (The first Halibrand magnesium wheels had the same Rudge-type splines as were used with wire wheels. Subsequently Halibrand changed to the familiar (and cheaper) pin-drive style.)

For the 1948 Indianapolis race, Kurtis also built a new car for Les Anderson of Tacoma, Washington. A long, rather ungainly vehicle, it was quite conventional aside from its tube frame. Anderson finished 24th at Indianapolis and was killed in another car at Portland, Washington, a month later. The car would go on to run in the East for two years with NASCAR's Speedway Division for Johnny Tadlock, with a DeSoto Hemi V-8 engine.

The *Kurtis-Kraft Special* Number 7, the KK1000 model, Frank Kurtis' first ground-up Indianapolis car that was destined to win the 500 in 1950. In five 500s it finished 9th, 2nd, 1st, 2nd, and 10th. *Arlen Kurtis collection*

KK1000 THE KURTIS-KRAFT SPECIAL

The Kurtis Indianapolis car that had been germinating in Frank's head for more than a year got to Indianapolis in 1948. Although Kurtis himself was an artist capable of designing a frame, sketching attractive bodies and shaping them in aluminum, he was not a trained draftsman, and he hired John Bond to do the engineering drawings of the KK1000. Dubbed the *Kurtis-Kraft Special*, the car featured independent suspension on all four wheels, using beefed-up Ford passenger car A-arms in front and a de Dion rear. It had a new rubberized fuel tank designed not to split in a crash.

Walt Brown, a journeyman driver who had placed seventh in AAA standings in 1947, was hired to drive it. Brown wrung the car out in practice at an early season Arlington Downs, Texas, 100-mile race and a part in the rear suspension broke. Kurtis hurried the car back to Glendale and replaced the casting with a fabricated steel part.

At Indianapolis, unfamiliar with anything but the old cross-spring suspension and possibly worried about a repeat of the failure in Texas, Brown was uncomfortable in the car and could not get it up to qualifying speed. Kurtis turned to Tommy Hinnershitz, another dirt track driver who was not thought to be an expert on pavement. Hinnershitz, a far more talented and analytical driver, adapted to the driving style the independent suspension required, qualified 14th fastest and was running as high as 4th place when the magneto died. Tommy limped to a stop, and the magneto was changed. The car's total time in the pits including the magneto repair was 24 minutes. Hinnershitz returned to the race to finish 9th, two laps down. The team took home $4,270, of which Hinnershitz' share was $1,700.

By 1948 the hoopla and excitement at Indianapolis had already begun to bother the very private Kurtis, just when he was beginning his years of success there. His custom Buick drew a crowd all around town so, after the race, Kurtis asked the team mechanic, Harry Stephens, to take the Buick while he drove the pickup truck that towed the race car. The entourage headed up the road to Milwaukee where, on June 6, a reinvigorated Brown qualified 10th on the dirt and finished 13th.

On June 20 they drove east to Langhorne, Pennsylvania, where Brown won, going away. After that, Kurtis left the race car in Stephens' hands and headed home.

The *Kurtis-Kraft Special* raced most of the Championship Trail that season. Harry Stephens put West Coast charger Johnnie Parsons in the car at Springfield, where he finished second. Johnnie repeated with another second place at Milwaukee August 29. After two so-so finishes at DuQuoin and Springfield, Parsons won the tragic October 10 race at DuQuoin where Ted Horn was killed.

Tommy Hinnershitz, the eastern dirt track star, stands by his *Kurtis-Kraft Special* Number 7. Despite a lengthy pit stop to replace the magneto, Tommy finished ninth. *Mark Dees collection*

Parsons' successes won him the ride in the *Kurtis-Kraft Special* for 1949 and he roared through that season. Johnnie took the opening race at Arlington Downs on April 24 and, in his second year at Indy, finished second to Bill Holland, who was driving one of Lou Moore's Blue Crown front-wheel-drive cars. Kurtis maintained afterward that Parsons could have won the race if pushed harder by his pit crew.

Bill Taylor gets advice before going out to try to qualify in the 1949 *Rounds Rocket*, the second rear-engine car to run at Indianapolis. Kurtis did not build the car, but he made major changes in it for Jack Rounds, but even so, Taylor could not get it up to speed. *Author's collection*

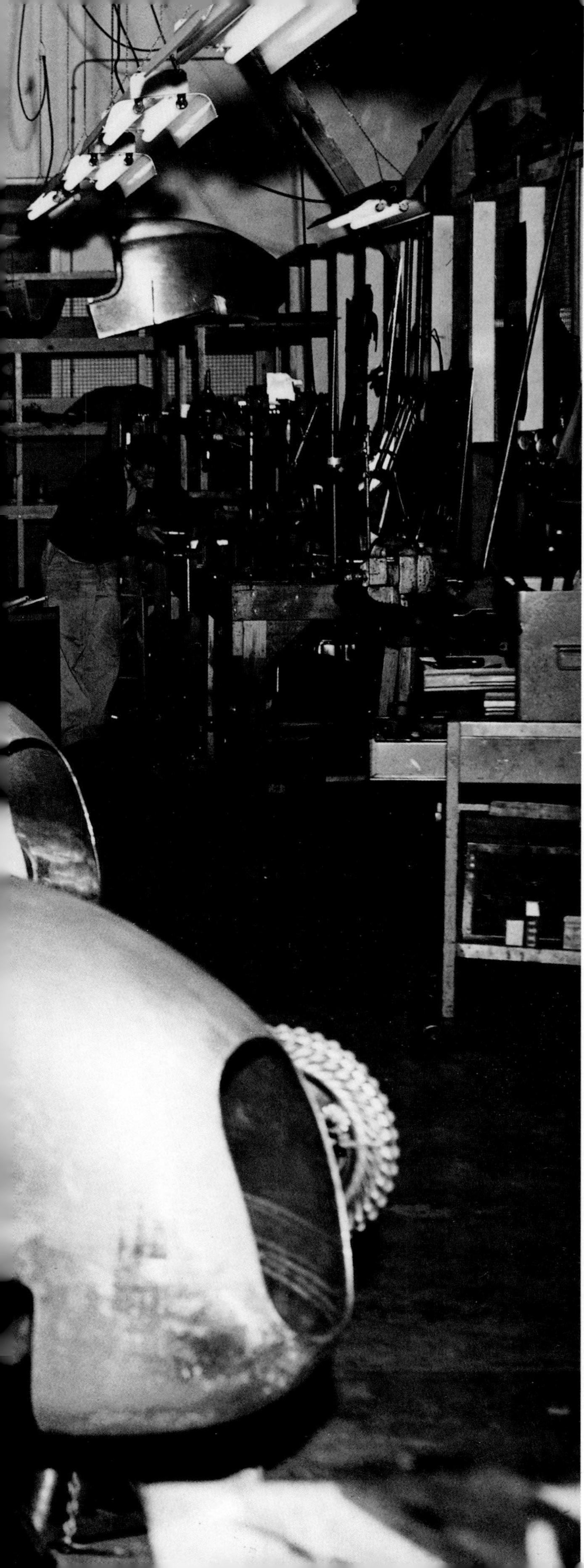

The Kurtis shop on North Alger Street in 1948, with four KK2000s under construction. The car in the foreground is the Agajanian Number 98 that Johnny Mantz drove at Indianapolis in 1948 and 1949. Note that these cars had onboard starters on their Offenhauser 270 engines. In many ways these were stretched midgets. In fact, they used midget tail stampings with a 5-inch section welded in. *Frank Kurtis*

Joe Scopa works on the Merz Engineering Number 23, a KK3000 that Sam Hanks drove in 1950. An oil leak put the car out and it finished 30th. *Author's collection*

KURTIS SPRINT CARS

Despite building hundreds of midgets and more than 100 Championship cars, Kurtis made few sprint cars. He explained in a 1978 letter, "We only built about a dozen sprint cars in our time. Sprinters didn't pay off much in those days and we had to get as much for them as we did for the championship cars. I guess they were too expensive for the customers." In 1980 he estimated that he built only six of the shorter wheelbase dirt track sprint cars. Historians, and his son Arlen, think only three Kurtis chassis were built specifically for the half-mile tracks, though several champ cars were cut down and many midgets "stretched" by later owners.

Gordon Reid in the *Engle-Stanko Special* Number 4 at St. Paul, Minnesota, in 1950. This was a shortened version of the Kurtis KK2000 car for half-mile tracks, and only a few were built. Reid was killed in the car at Dayton, Ohio, April 20, 1952. *Kem Robertson collection*

The first B.C.R.A. sanctioned sprint car race was at the Colorado State Fairgrounds in Pueblo, CO. in 1958. Wayne Weiler in the Fike Plumming Chevy and Al Sherman in his stretched Midget, the "Jewel Box" each won a Feature during the two day show. Al later raced the "Jewel Box" until the mid 1960s on the Midwest I.M.C.A. circuit.

He did build some short-wheelbase versions of the KK2000 dirt championship cars, notably cars for Leo Dobry, Hal Cole, and J. C. Agajanian, in 1948. All three, in fact, ran at Indianapolis as well as on the dirt. There is some indication that the wheelbase was stretched for Indianapolis and shortened for the mile and half-mile dirt tracks.

Three cars built with the 88-inch wheelbase of half-mile cars have been identified in the Kurtis records, for Charles Engle, Mari George, and Jim Campbell. The Engle-Stanko Number 4 was a shortened, lightened KK2000, while the Mari George sprinter was a stretched version of the Kurtis midget. Jim Campbell's IMCA sprinter was also just a midget frame lengthened by 16 inches. Kurtis said in 1982 that Engle "originally ordered a KK2000 series (car) but after we got started, he changed his mind and we made a shorter-wheelbase sprint car."

Off the Kurtis books was Johnnie Parson's *AutoBrite Special* Number 4, another stretched midget frame carrying a 220-cubic-inch Offy. (Notable Kurtis midgets that were later stretched into sprint cars included Pike Green's *L'il Stinker* Number 96 and Mike Caruso's Number 36, with its supercharged midget engine.)

There are reports that two or three other sprint cars were built on midget frames by Kurtis employees on their own time, but they were not recorded on the Kurtis-Kraft rolls. A number of sprint cars built elsewhere used Kurtis-supplied aluminum noses and tails.

Leo Dobry's *City of Tacoma Special* Number 63, a KK2000, with the tail off showing the rubber-covered fuel tank and the aircraft-style brakes. Hal Cole drove the car, powered by a 220 Sparks engine, to sixth place in 1948. *Author's collection*

and widened to accommodate the standard 270-cubic-inch Offenhauser engine, 1-1/4 inches wider than the Offy midget. The KK2000s are distinctive because of their hairpin radius rods and semicircular radius rod mounts on each side of the car. Hard to change later, these are prime identifying features.

Because of the longer frame of the 2000, Kurtis used two parallel 1-3/4-inch-diameter tubes with a flat strip of steel welded between them for stiffness. Most were delivered with parallel torsion bar suspension; Conze quick-change rear axles; Houdaille rotary-vane hydraulic shock absorbers; Rudge 52-millimeter, 16-inch wire-spoke wheels; Ross steering; aircraft brakes; 60-gallon fuel tanks; and a 270-cubic-inch Offenhauser engine with an on-board starter. The basic chassis, less engine, sold for $6,736, or

$8,000 with paint, chrome, and upholstery. The first 2000, sold to Leo Dobry of Seattle, came with an overhauled 247-cubic-inch Sparks version of a Miller four-cylinder engine.

Once word got out that Kurtis was building a conventional Championship car that promised to be as successful as his all-conquering midgets, customers lined up. Five 2000s were built for the 1948 race. Another eight were built in 1949, though two of those were sold as incomplete cars to George Tuffanelli, of Chicago, who had Charles Pritchard, Tudy Marchese, and Joe Silnes modify them with extra shock absorbers, straight axles, and Ford-style radius rods, including a change in the radius rod mounts.

They were not as successful at Indianapolis as the Kurtis midgets were on the short tracks. In their first three years, when they were the most competitive, (1948, 1949 and 1950) the 2000s managed only six top-10 spots at Indy. Their best placings were Joie Chitwood's fifth place finishes in 1949 and 1950. It is said Chitwood could have won the race in 1949, except that the grille came loose when it was hit by debris from Charlie Van Acker's crash. The time wasted in several pit stops to deal with the grille cost Chitwood the difference between winning and losing.

In general, the KK2000s at Indianapolis were about 3 miles an hour slower over the 500 miles than the winning cars in 1948, 1949, and 1950. The KK2000's shining moment came in 1950 when, on May 13, Walt Faulkner, the "Little Dynamo," driving J. C. Agajanian's two-year-old *Grant Piston Ring Special* Number 98, qualified at 134.343 miles per hour, breaking the track record set by Ralph Hepburn in the Novi four years earlier. The Number 98, wrenched by Clay Smith, was the most successful KK2000. Not only did it sit on the pole at Indy in 1950, but it won AAA 100-mile races at Milwaukee and Sacramento and a 200-miler at Milwaukee.

After 1950 the KK2000s were second-tier cars at Indianapolis, run by struggling, underfinanced teams or those

The Agajanian Number 98, a 1948 KK2000, after Walt Faulkner, the "Little Dynamo" put it on the pole at Indianapolis in 1950 with a sparkling 134.343–mile per hour qualifying run. Faulkner came in seventh in the race. *Author's collection*

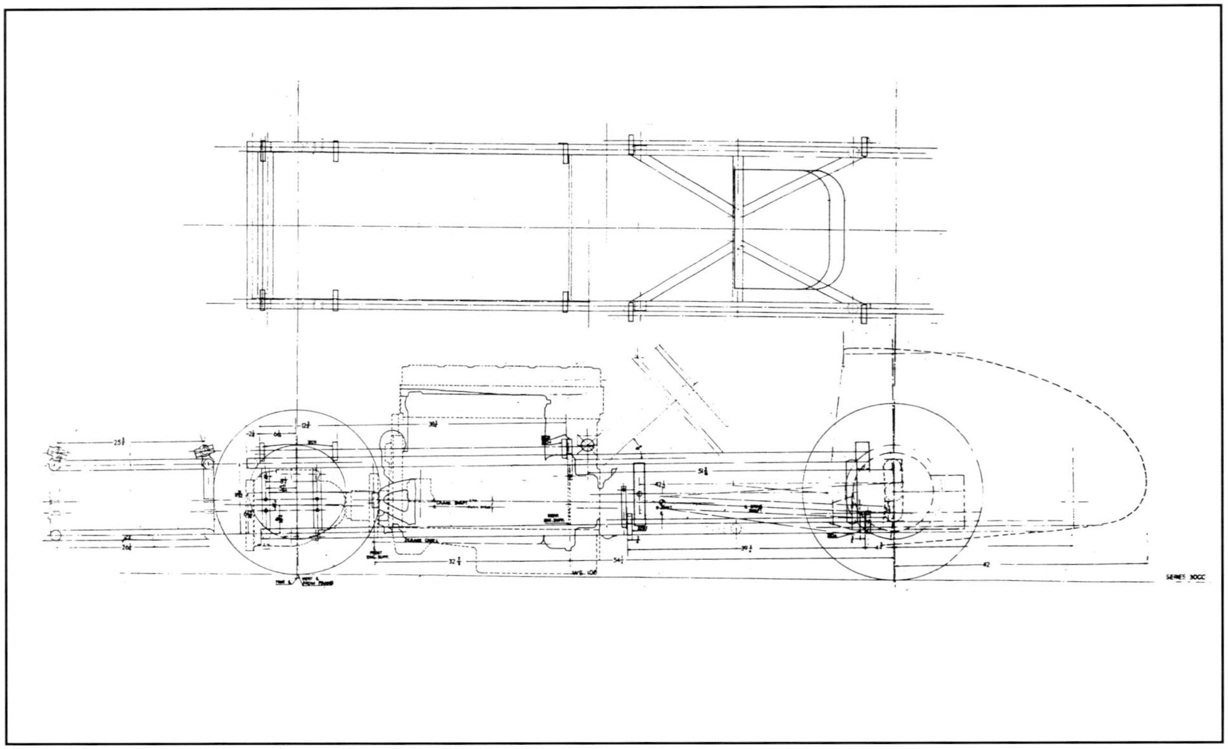

Chassis layout of the Kurtis KK3000 Indianapolis car, done by Carl Newell, in 1950. *Leo Goossen Archive*

Jack McGrath in Jack Hinkle's KK3000 at Indianapolis in 1951. Hinkle finished third in this car. *Tom Saal collection*

The McNamara number 73, a KK2000, at Indianapolis in 1953 when driver Carl Scarborough collapsed of heat exhaustion and died. Relief driver Bob Scott brought the car home 12th. *Robert McConnell collection*

content to compete without expecting to win the race. The last to qualify there, in 1955, was Sam Traylor III's Number 42, driven by Al Keller and maintained by Ken Hickey.

On the dirt tracks, the KK2000s were more successful. Owners scrapped the aircraft brakes and often put a Ford style cross-spring in front and doubled up the shock absorbers. Over five years they won six 100-mile AAA dirt track Championship races.

THE PIERSON CARS

In late 1949 and early 1950, Gil Pierson, designer of the racing fuel pump that later became the standard Hilborn unit, ordered two front-wheel-drive cars that looked for all the world like miniature versions of Lou Moore's Blue Crown cars. Leo Goossen drew up the designs and Kurtis built them. For reasons not understood today, neither car was ever entered at Indianapolis or anywhere else as far as is known. One of them stood in for a Blue Crown car in a movie role, but they never raced competitively. A set of spares for the front-drive transaxle found their way to Gordon Schroeder, and Emil Deidt built a sleek aluminum frame car around them, the *Tuffanelli-Derrico* Number 19 that Mack Hellings drove at Indianapolis in 1951.

THE KK3000

In 1950, building on success of the *Kurtis-Kraft Special*, Kurtis designed a car to replace the KK2000. Designed primarily for Indianapolis, he called it his model KK3000. Using what he had learned from the house car, it had independent front suspension using Ford parts, a solid open tube Halibrand rear axle, and parallel torsion bar springs with hairpin style radius rods. The KK3000s weighed 1,650 pounds and carried between 53 and, with an auxiliary tank, 72 gallons of fuel. Ready to race, they sold for about $20,000.

Bill Holland in the *Slick Racers* Number 61, a KK3000, in 1953. This was by then a four-year-old car. Jack McGrath drove it from 1950 to 1952 for Jack Hinkle, with a third place in 1951 their best. Even though Holland bought a new 270 Offy engine for it, he could not get it up to qualifying speed. Ed Stone and Duke Donaldson tried with it for three more years without success. *Author's collection*

Cecil Green in the *John Zink Special* Number **4**, a KK3000, in 1950. Green finished fourth in this M. A. Walker entry. *IMS*

Kurtis himself entered the first KK3000 at Indianapolis in 1950 as the *Wynns Friction-Proofing* Number 28, with Freddie Agabashian as his driver. It finished 28th when its supercharged 177-cubic-inch Offy broke its special blower drive gears. Kurtis sold the car, spares, trailer, and tow truck to Ed Walsh for $15,000.

CUMMINS DIESEL OF 1950

The Cummins Engine Company approached Kurtis in 1950 for a chassis in which to run its 400-cubic-inch truck diesel. Clessie Cummins had run diesel cars at Indianapolis sporadically since 1931. The 1950 Cummins Indianapolis engine was placed upright in a modified KK3000 chassis; it qualified 32nd, going out on Lap 52 with a clogged supercharger.

Six other KK3000s were built, in 1950 and 1951, for M. A. Walker, Andy Granatelli, Jack Hinkle, Dan Levine, Eugene Cassaroll, and Peter Schmidt. The type was moderately successful at Indianapolis with 10 top-10 finishes in five years, the best being Jim Rathmann's 2nd place in 1952 in the ex-Granatelli car, then owned by Ed Walsh and sponsored by Wynns. They faded fairly quickly from the Indianapolis scene, with the last appearing in the race as Chuck Weyant's *Federal Engineering* Number 41 in 1955, which finished 12th. No KK3000 ever sat on the pole at Indianapolis. With their independent front suspensions, they were not designed for dirt and in fact won only three races in their rather brief four-year competitive span. Jack McGrath won at Langhorne and Syracuse in 1950 in *The Hinkle* Number 49 and Louis Unser won Pikes Peak in 1953 in *Levine's Federal Engineering* Number 7, probably more a tribute to Unser, the Old Man of the Mountain, than to any special attribute of the KK3000.

The 1952 Granatelli KK3000 that Rathman drove had been used before that year's Indianapolis race as a tire test car for Firestone with a Cadillac V-8 engine. Photos of it in race trim with a 270 Offy engine show the cutouts under the hood where the Cadillac exhaust pipes exited. Granatelli had hoped to get AAA to allow him to run the Cadillac stock block engine at 330 cubic inches, but of course the officials laughed off that idea. Big stock blocks would reappear later in the year in the form of tire tests with 330-cubic-inch Chrysler Hemis.

THE NUMBER 99

By far the most successful Championship car to come out of the Kurtis-Kraft plant was a 1949 collaboration between Kurtis and Louis Meyer and Dale Drake, then builders of the Offenhauser engine. Meyer & Drake thought a supercharged engine would outclass normally aspirated engines under the existing rule that gave the supercharged powerplant two-thirds the displacement of unsupercharged engines. A blower can easily pump in twice the charge that atmospheric pressure can get through an engine's valves in the brief time that they are open. Meyer & Drake had tried adapting a blower to smaller versions of their engines and liked what they saw on the dynamometer. Leo Goossen designed a shock-absorbing coupling that appeared to solve the blower's Achilles' heel of broken driveshafts.

They put together a centrifugally blown version of their midget engine and had Kurtis stretch one of his 72-inch-wheelbase midget frames to the Championship length of 96 inches. The parallel torsion bars were mounted relatively high in the frame and the car had solid axles fore and aft, in good sprint car style. It was relatively rigid and quite light, at 1,340 pounds. When it was finished in August, most of the M & D staff posed with the red Number 99 car in front of their shop, while three-time Indianapolis winner Lou Meyer sat behind the wheel.

The cockpit of the M. A. Walker Number **54**, a 1950 Kurtis KK3000 Championship car that Cecil Green drove to 4th place at Indianapolis in 1950. Green finished 22nd in 1951 with a broken rod, and even tough Mike Nazaruk could not get it up to qualifying speed in 1952. *LeRoy Byers*

Don Cummins and Frank Kurtis with Jimmy Jackson in the 1950 *Cummins Diesel Special* Number 61 and Johnnie Parsons in the *Kurtis Kraft Special* Number 1 in 1950, the year Parsons won the 500. *Cummins Engine Co.*

The 1950 *Cummins Diesel Special* Number 61, a KK3000 chassis with the hood raised to accommodate the upright truck engine of 400 cubic inches. *Arlen Kurtis collection*

Meyer asked Indiana Chrysler dealer Murrell Belanger to test the car on the Midwestern dirt tracks. Mechanics were his son, Lou "Sonny" Meyer Jr.; George Salih; Chickie Hirashima; Frenchy Sirois; and Belanger's Tiny Worley. Belanger put Tony Bettenhausen in the car. Bettenhausen, a midget driver, had some champ car experience, having won the 100-mile races at Goshen, New York; Springfield, Illinois; and, in relief, at Milwaukee. After dropping out at Springfield and failing to finish at Milwaukee that summer of 1949, their next race, at DuQuoin, Illinois, was a runaway. They moved on to Detroit a week later and won again. Rumors quickly circulated that Meyer & Drake had some engine secrets they were keeping to themselves and not letting their customers have. In the face of such complaints, Meyer & Drake stood the car down for the rest of the 1949 season.

Over the winter they arranged to sell the Number 99 car to Belanger, who repainted it a deep blue with glistening gold leaf numbers.

Belanger entered the supercharged car at Indianapolis in 1950, but it was no world-beater. Neither second-year man Kenny Eaton nor old-timer Emil Andres could put it in the race. After an unhappy month of May, Belanger pulled out the blown midget engine and put in a standard 270 Offy. Chuck Stevenson put it on the pole at Milwaukee in August 1950, but wrecked it in a north turn crash. Neither Duane Carter nor Harry Turner did much with it on the dirt tracks that summer. Belanger put Bettenhausen back in the driver's seat in September. Bettenhausen and the Number 99 were the

Jimmy Jackson in the 1950 *Cummins Diesel Special* Number 61 at Bonneville, where he ran 165.20 miles per hour, setting a diesel record. This was a KK3000 model with a hood line raised to accommodate the huge Cummins engine. Jackson had finished 29th in the car at Indianapolis in 1950. Pieces of it went into Cal Connell's Cadillac-powered Indy car. *Author's collection*

Vince Granatelli works on the Cadillac V-8 in the *Grancor-Wynns Special* Number 59 in 1952, a KK3000. Andy had hopes he could run a big stock block at Indy. When he could not, he replaced the Cadillac with an Offy, and Jim Rathmann brought the car in second behind Troy Ruttman. In photos of the Rathmann car, the cutouts for the Cadillac exhaust can be clearly seen. *Author's collection*

Lee Wallard in the Belanger Number 99, the most successful Championship car Kurtis built, winner at Indianapolis in 1951 and in fifteen 100- and 200-mile races on the dirt. *Author collection*

magic combination. Bettenhausen won with the car at Springfield, was second at Phoenix, and finished the season winning at Bay Meadows.

In 1951 at Indianapolis, Bettenhausen was committed to driving the same Lou Moore Blue Crown car he had piloted in 1950, the car that had won the 500 in 1948 and 1949. Despite his successes on the dirt tracks with the Number 99, Tony had a deal with Moore, and in his mind the Blue Crown (by now sponsored by Mobilgas and owned by Bill Ansted) was a better bet at Indianapolis than a dirt track stretched midget that had not even qualified a year earlier. Belanger hired eastern dirt track driver Lee Wallard, a four-year veteran with two top-10 Indy finishes, to drive the Number 99.

Wallard found the dark blue Belanger car to be a dream on the Indianapolis bricks. In qualifying in 1951 it carried a 263-cubic-inch Offenhauser engine, unsupercharged. Lee came in from a fast practice run saying "Wot a car!"—a slogan Belanger put up as a banner over his garage in Gasoline Alley. The Number 99 started the 1951 race in the middle of the front row. It had used a steep rear axle ratio to qualify and because at that time a car was required to race with the same ratio it used to qualify, but not the same engine, Worley installed a light shorter-stroke 220 Offenhauser, bored and stroked to 241 cubic inches, for race day.

In the race Wallard and the Number 99 walked away with the victory, while Bettenhausen's obsolete ex–Blue Crown front-wheel-drive car finished ninth. Indianapolis was the fifth victory for the Belanger Kurtis on the AAA Championship Trail. Sixteen Kurtis chassis had qualified and, with Nazaruk and McGrath coming in second and third, Kurtis cars swept the top three places.

Bettenhausen became Belanger's driver of the Number 99 again after Indianapolis as, tragically, Wallard was

The Belanger Number 99, the winningest Kurtis Championship car, at Indianapolis in 1951, when Lee Wallard won the 500. Tony Bettenhausen made the car a legend on the dirt mile tracks during the rest of the 1951 season, after Wallard was badly burned in another car. *IMS*

The Kurtis shop crew and the *Kurtis-Kraft Special* Number 7, after it was completed in the early spring of 1948. Left to right are Joe Ward, Carl Blackman, Rod Schapele, El Johnson, Don Allen, Quinn Epperly, Tom Barnes, Tony Coleman, and Lou Salzgaber. Harry Stephens, the chief mechanic, is seated in the car, and Frank Kurtis is beside him. *Mark L. Dees collection*

The parade lap for the 1951 500, with Duke Nalon in the Kurtis-built Novi on the pole. Lee Wallard is beside him in the Belanger Number 99, another Kurtis, the Number Two starter. *Author's collection*

severely burned a few days later in a sprint car at Reading, Pennsylvania, and was not able to drive for the rest of the year. In fact, the Reading fire was a career-ending accident for Wallard, who tested at Indianapolis in 1952 and 1954, but found he was no longer capable of competitive driving.

But 1951 was Bettenhausen's and the Belanger Kurtis' year. Seldom was a combination of car, driver, and mechanics more successful. They won the 100-mile dirt track race at Milwaukee a week after Indianapolis and simply demolished the Championship Trail, taking 8 of 14 races that season, plus seconds at Darlington, South Carolina, and Williams Grove, Pennsylvania.

In 1951, Bill Schindler drove the Chapman Number 17 (the Kurtis 2000 in which Rex Mays had been killed in 1949). At Langhorne, Pennsylvania, Schindler took an early lead and held it to well past the 50-mile point, whereupon Bettenhausen in the Number 99 took aim on him and simply motored easily past for the win. The same scenario played out at Milwaukee, where Cecil Green led for 63 laps until Bettenhausen in the Number 99 decided it was time to take charge, and he did. The Number 99 car was, that year, almost unbeatable. When Bettenhausen lost, it was for a blown tire, a tangle with another car, or because it ran out of fuel.

As the field roared into the first turn at Indianapolis in 1951 Jack McGrath in a KK3000 took the lead from Duke Nalon in the Kurtis-built Novi, with Lee Wallard in the Number 99 Kurtis taking second place, followed by Troy Ruttman in the Number 98 KK2000. Tony Bettenhausen in one of the Blue Crown cars was next, followed by Mauri Rose and Duane Carter in the other two Blue Crowns. *Author's collection*

Lee Wallard steps out of the Belanger number 99 Kurtis after winning the Indianapolis 500 in 1951. This car was the second Indy-winner to come from the Kurtis shop. *Bob Verlin*

But Bettenhausen seemed jinxed at Indianapolis, and the gremlin struck early in 1952 as he spun the Number 99 into the wall on a qualifying attempt, damaging the car's front end too seriously to allow another try. Bettenhausen again took over the old Lou Moore Blue Crown front drive, then owned by Earl Slick, and finished 24th when, after the engine stalled in a pit stop, the starter shaft jammed.

Two exact copies of the Number 99 were made, one by Lujie Lesovsky for Belanger, and one by Kurtis that

The Belanger Number 99 after Tony Bettenhausen's crash during a qualifying attempt in 1952. Cal Niday and Jerry Hoyt qualified the Number 99 in 1953 and 1954. The car never finished higher than 26th at the Speedway after 1951, though with Bettenhausen it was dominant on the dirt tracks. *Author's collection*

Johnny McDowell bought. Neither did particularly well. The Belanger copy finished 4th in 1952 and 24th in 1953 for Duane Carter. McDowell finished 21st at Indianapolis in 1952 and died in a practice crash at Milwaukee a week later. The Number 99 is recorded as did not qualify at Milwaukee that day, but it was McDowell's death, not lack of speed, that kept Bettenhausen out of the race. Shocked by his death, Bettenhausen decided to retire from racing except for Indianapolis.

Without Bettenhausen, some of the magic was gone for the Belanger team, but the Number 99 was still quick. Jim Rathmann drove it to a respectable 4th at Raleigh, North Carolina, on July 4, but he couldn't qualify at either Springfield or in the August 24 race at Milwaukee. Worley, who was running the Belanger team, replaced Rathmann with Paul Russo, Bettenhausen's old buddy. They placed 2nd at Detroit, 4th at DuQuoin and at Syracuse, but among the back markers at Denver and Phoenix. At Indianapolis in 1953, Cal Niday qualified 30th in the car, by then called the *Belanger Miracle Power Special*, and finished 30th. On the dirt, Niday and Jimmie Davies drove the car, with Niday's 4th at Springfield their best showing.

By September 1953, Bettenhausen could no longer bear to be away from racing. He called Belanger and took the seat in the Number 99 back from Niday for the Labor Day race at DuQuoin, Illinois. Still rusty, Bettenhausen qualified 6th and went out on the 85th lap with a rear axle oil leak. But five days later at Syracuse, New York, he and the car were back in top form. Bettenhausen and the Number 99 won by nearly two laps over Mike Nazaruk. At Phoenix on November 11, Bettenhausen and the Number 99 were first again, a lap ahead of Duane Carter in the *Auto Shippers* Number 23, the 15th checkered flag for the Number 99.

In 1954 the invasion of the roadsters was in full swing at Indianapolis. Bettenhausen had a ride in a new Kurtis 4000 owned by Mel Wiggers, an upright dirt track car, but more modern than the five-year-old Number 99. Belanger put Jerry Hoyt, a fourth-year man, in the by-then venerable car. Hoyt qualified it 30th and finished 26th when the engine blew on the 130th lap. Hoyt could not qualify at Milwaukee in June; he got a 5th at Darlington, South Carolina, but couldn't make the show at Springfield, Illinois, or the August race at Milwaukee. Belanger replaced him with Sam Hanks, who showed that the Number 99 was still a champion by winning the Labor Day race at DuQuoin. Hanks and the Number 99 were a strong 3rd at the Indianapolis Fairgrounds September 18, ran 5th at Phoenix, but did not qualify at Las Vegas in November.

In 1955 the Number 99 car was no longer entered at Indianapolis in May. Belanger was cutting back his racing, but took the Number 99 to Milwaukee August 28, where eastern newcomer Eddie Sachs tried to qualify it, but was too slow. Johnnie Parsons, on his way to a graceful retirement, failed to get it up to speed at DuQuoin in September.

On September 17, 1955, the Championship cars were at the Indianapolis Fairgrounds, a 1-mile dirt track, for their last appearance east of the Rocky Mountains under the AAA banner. Vukovich's death in May and the Le Mans tragedy had led the American Automobile Association to abolish its Contest Board and drop out of any connection with auto racing. Belanger brought the Number 99, and Bettenhausen would drive it for the last time. They qualified seventh and finished fourth. Not bad for a car that had rolled out of the Kurtis plant six years earlier. Bettenhausen had driven the car for 14 of its 16 wins.

Was the lightweight stretched midget design of the Number 99 something special? Neither Johnny McDowell's nor Belanger's copies did well. After McDowell died in the car at Milwaukee, Roger Walcott, his sponsor, hauled the wreckage back to Indianapolis, and over the winter of 1953 built a dirt sprint car out of the parts in partnership with Mari Hulman. This car became Hulman's H-O-W Number 55.

It would seem that the design itself, even when run by the same owner, was not successful without Tony Bettenhausen as the key ingredient. Likewise Bettenhausen was not as successful as a driver in a car other than the Number 99. There was a synergy between car and driver that lifted both to Championship caliber.

KK4000

The KK4000, brought out in 1951, was the last of the Kurtis Championship chassis to be designed with both Indianapolis and the dirt tracks in mind. More than a slightly enlarged KK2000, itself a bigger version of the Kurtis midget, the 4000 frame was built of an oval tube of perforated plate with tubing hoops to support the body, as in the midgets. It rode on conventional solid axles suspended with parallel torsion bars contained in the tubes. Kurtis was not ready to try his still-percolating offset design on the dirt, where getting a "bite" required

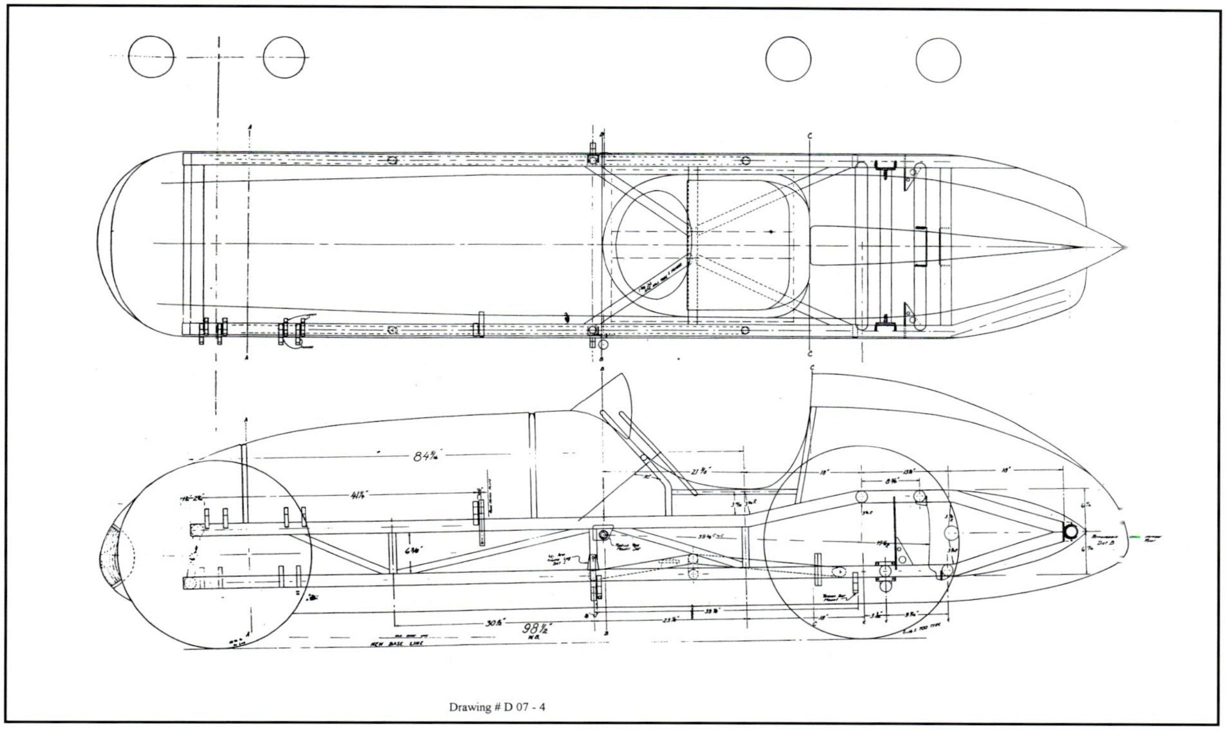

Chassis layout for the Kurtis KK4000 car, basically a dirt Championship car, but one that ran at Indianapolis from 1951 until 1956. *Leo Goossen Archive*

Ed Walsh's two KK4000 Bardahl-sponsored cars; Jimmy Davies in Number 45 and Jim Rathmann in Number 1 dice for position at Indianapolis in 1954. *Armin Krueger*

a different approach than with a pavement car, so the 4000 was a traditional upright, as are dirt track cars even into the twenty-first century.

The first KK4000 went to auto dealer Hart Fullerton, who put a blown 183-cubic-inch Offy into it. Bill Cantrell was not able to get it up to qualifying speed at Indianapolis in 1951, but in 1952 Andy Linden put it in the field at better than 137 miles per hour, second-fastest on Pole Day to the Cummins Diesel, but Linden went out of the race with an oil leak on Lap 20.

Fullerton ran his KK4000 at Indianapolis through 1956, and then campaigned it in dirt track races until 1962, when Andy Granatelli bought it as a training vehicle for Art Malone. Malone, a drag racer, had been chosen by Granatelli to drive his new Novi, and the USAC officials told him to get some oval track experience before attempting Indianapolis.

Fifteen KK4000s were built between 1951 and 1953. They were considered primarily dirt track cars, and none but the Fullerton car ran at the Speedway their first year. They were entered in 1952 and their best finishes at Indianapolis came in 1953, when Art Cross was second in Ray Brady's car and Sam Hanks was third in Ed Walsh's *Bardahl Special*. Most of the 4000s eventually did run at the Speedway, and they did fairly well on the pavement surface they were not really designed for.

The best-known of the KK4000s was probably the *Springfield Welding Special* Number 16. (Serial Number 349) Chuck Stevenson drove it to victories at Milwaukee and DuQuoin and finished well enough on the rest of the Championship Trail to take the 1952 AAA National Championship.

Rodger Ward drove M. A. Walker's KK4000 (Serial Number 340) in 1953 and 1954 on the dirt, and he won his first Championship race in it at Springfield, Illinois, June 21, 1953. The M. A. Walker Number 92 car was actually something of a hybrid between the KK3000 and KK4000, with the wider body of the 3000 and the solid axle suspension of the 4000.

Sandy Belond bought the last 4000 in 1953 (Serial Number 368) for Johnnie Parsons to drive in dirt 100-milers. This was a year after the KK500 roadster had appeared at Indy and Belond, who also bought a Kurtis roadster for the Speedway, had no thought of running the 4000 there. Bruce Bromme campaigned the car for Belond with Parsons in 9 of the 12 AAA dirt track races in 1953, with a second at Williams Grove, Pennsylvania, its best finish. Belond apparently parked the car in 1954 and 1955. In the fall of 1955 he sold it to Sam Traylor III of Allentown, Pennsylvania, shipping it east by rail. The car was "lost" in the rail yards in St. Louis for a month but finally made it to Ken Hickey's shop north of Philadelphia.

Hickey, a consummate race mechanic, rebuilt the car and Traylor's year-old 270 Offy engine. In 1956 Al Keller qualified the Traylor Number 55 at 141.193 miles per hour, the fastest a dirt car ever ran at the Speedway and the last.

LAST KURTIS CHAMP CAR VICTORY

Traylor ran the car at Langhorne, Trenton, Williams Grove, and other eastern tracks for three years, then sold it to Jake Vargo, another easterner. Under Vargo's ownership it won the last Championship victory for a Kurtis-built car, with Van Johnson driving at Langhorne, Pennsylvania, on June 14, 1959.

The ex-Belond car had a tragic history under Vargo. Dick Linder died in the car at Trenton in April 1959. A month after his Langhorne victory, Van Johnson was killed in it at Williams Grove. At Langhorne, on July 1, 1962, Bob Mathouser qualified the Vargo Number 63 as seventh, but pulled in on the 33rd lap, exhausted from manhandling the old and ill-handling car. Mathouser, in turn, was relieved by Elmer George, who took it for 16 laps but then he gave it up. Hugh Randall was there in the Ray Brady Number 46, another KK4000 that was too slow to qualify. When George quit the Vargo car, Randall took it over in relief. Two laps later, he crashed the car and died.

Kurtis 4000s ran longer than most of the other Kurtis Championship designs, many on through the 1960s. Carl Gelhausen ran his *Midwest Manufacturing* Number 84 until at least October 4, 1970, when Tom Bigelow finished 17th in it at Sacramento. The car had been bought 19 years earlier by John McDaniel as the *Blakely Oil Special*.

Johnnie Parsons at Williams Grove, Pennsylvania, in the Belond KK4000 dirt car in 1953. This car would go on to be owned by Sam Traylor and Jake Vargo and to win the last championship race by a Kurtis chassis. *Bruce Craig*

THE 1952 CUMMINS DIESEL

In the early 1950s, the Cummins Engine Company continued its interest in racing. The standard Cummins truck diesel at that time displaced 400 cubic inches, nearly 50 percent more than allowed to the normally aspirated Offenhauser engine. The diesel engine size limit in 1934 had been 366 inches. By 1950 the rules had somehow been liberalized, and at 401 cubic inches, the turbo-supercharged diesel was competitive.

Kurtis had been advocating offset and laidover engines for at least four years when he built the 1950 Cummins, and the tall truck engine was a prime candidate for the scheme. In 1951, unsuccessful in trying to sell his laydown design to the Indianapolis establishment, Kurtis was ready, he said, to give it up and concentrate on work for the aircraft companies and custom car buyers. But he was soon to change his mind.

After the 1951 race, Don Cummins and several of his engineers huddled with Kurtis in Cummins' passenger car in the garage area while the rain pattered down, and

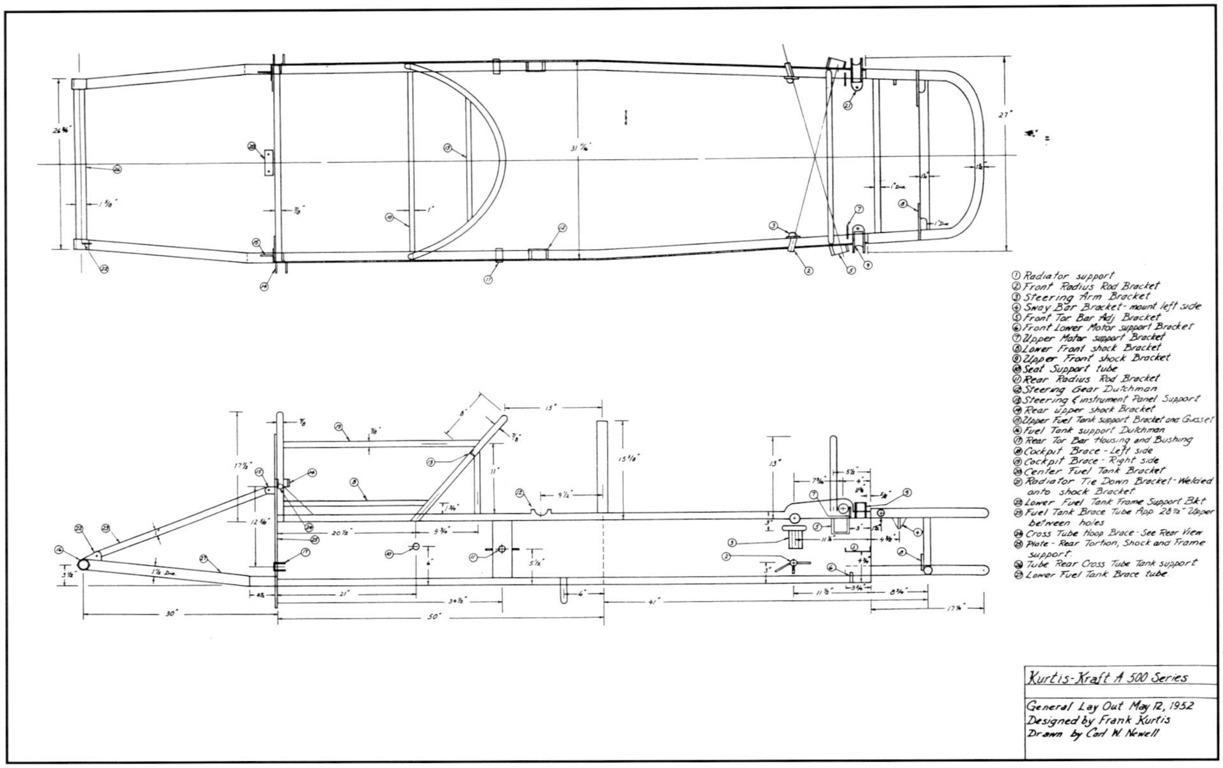

Chassis layout by Carl Newell of the KK500A, the first Indy roadster. *Leo Goossen Archive*

went over plans for a laydown version of the engine. Later Cummins took Freddie Agabashian, a solid, analytical driver recommended by Kurtis, to a quiet spot on the backstretch, out of sight of curious eyes, and unrolled a tentative sketch for the new diesel racing car. Agabashian was immediately impressed with the idea of a car carrying the engine laid over, reducing frontal area and allowing the driver to sit alongside the driveshaft, rather than over it.

Cummins put Agabashian on the payroll as a consulting engineer with a 12-month contract starting June 1, 1951. The blueprints, drawn by Kurtis and checked by Cummins engineers, were sent to Glendale, where Kurtis-Kraft started assembling the chassis in the fall.

Agabashian said the laydown Cummins "absolutely delighted" Kurtis.

"Frank had been getting weary of proposing new designs and having the race crowd say, `No, we've always done it this way, so just make a few little changes,'" Agabashian said. "He was on cloud nine, having someone enthusiastic about a new concept and be willing to pay for it."

The resulting car was built around a wooden mockup of the Cummins diesel, set at an angle of five degrees off horizontal—just enough so that oil would drain out of the cylinders when the engine was at rest. Agabashian went to Glendale three times to be measured and have the cockpit fit to his height, reach, and weight. In its suspension, it was a beefed-up KK3000 with independently sprung front wheels.

The first tests at Indianapolis came in late November 1951, and the car was then flown to Wichita, Kansas, just before Christmas for two days of wind tunnel work by Dr. George Rasak, an aerodynamicist at the University of Kansas. Cummins' willingness to pay for air freighting the Diesel between Los Angeles and Indianapolis amazed Kurtis, who was used to race cars being towed across country on open trailers behind the owner's station wagon. When he realized that if he delivered a car at Indianapolis rather than in Glendale he could avoid California state sales taxes, many of his Indy cars were shipped that way, and the tax savings paid for the air freight.

While the Sparks-Weirick *Catfish* of 1932 had been designed along aerodynamic lines suggested by Stanford University Instructor Ulysses A. Patchett, the Cummins car was the first Indianapolis car to actually be tested in a relatively sophisticated wind tunnel. It was by far the most scientifically advanced body design Frank Kurtis had built up to that time. Agabashian, who attended the wind tunnel tests, said they found that the original windshield took upward of 60 horsepower to drag through the wind. Dr. Rasak was able to trim that to 9 horsepower, but the driver had to keep his head and hands completely out of the 140-mile per hour slipstream. Finally, a compromise windshield design was worked out that allowed the driver some movement but took only 14 horsepower.

Other changes were worked out; for example, air that was allowed to enter the nose of the car created far more drag than having it pass around the body. Shutters were installed that allowed the driver to open and close the cooling air intakes. Agabashian said he would open and close the intakes at Indianapolis four times each lap and that the drag when opened was immense. The wind tunnel work also helped develop ways to direct air to all of the radiator rather than just the center. During the tests they discovered that vibrations thought to be caused by the rough Indianapolis bricks were actually aerodynamic flutter of some of the body panels.

Kurtis was delighted with the wind tunnel work. He learned things that he would soon use in building his next cars, the KK500 series. Agabashian was delighted too, with the car that was built for him. The one downside of the roadster style, with its more enclosed cockpit, was that the driver got less ventilation and was more bothered by engine heat. "But I loved it," Agabashian said years later. "It was my favorite race car. It was built for me. It fit like an old glove." When, on its 30th anniversary in 1982, the Diesel was run again at Indianapolis, Agabashian was there to drive it. "It was wonderful. It was my first love," he said when he climbed out of the car.

Despite the great weight and size of the engine, the Cummins car was fast, though it accelerated slowly compared to the Offenhauser-powered cars. In May, at Indianapolis, Agabashian smoothly brought the Cummins up to speed and, on the first day of qualifying went 138.010 miles per hour, breaking the track record and putting the Cummins on the pole. Later in the month, Chet Miller would drive the Kurtis-built Novi 139.034, and Billy Vukovich would drive the Keck 500 Kurtis 138.212, but neither speed was set on Pole Day. Aside from the Novi and the Keck cars, no one else that year was within a mile an hour of the diesel's speed.

Cal Connell's Kurtis-Cadillac of 1953. It had the independent front suspension of the KK3000 and the rear of the KK500. Some of its parts came from the 1950 Cummins Diesel, and Ed Gdula did the assembly in Detroit. Connell, builder of the Crusader Cadillac boat engines, had this one set up for reverse rotation, but the engine blew in a qualification attempt. *Bob Mount*

Before the race Agabashian, the pole driver, had asked Wilbur Shaw, who drove the pace car, to go as fast as possible on the pace lap so the Cummins would be nearly up to speed. "But Wilbur wanted a picture-perfect start," Agabashian said, "and it was not fast enough for me."

"I knew what the Diesel would do at the start, and I didn't want any tire tracks on my helmet as the hot dogs rode over me, so I moved over to let them go by. Between the starting line and the first turn, I was back to 9th and by the back stretch, I was 12th," he said. "It took that long to get the Diesel up to speed. But then the car had accelerated and in another lap I was back to 9th."

The Cummins car ran very well, though its great weight was hard on tires. The blower clogged with rubber particles, and the Diesel dropped out on its 71st lap. This was due, Kurtis said, to a bad design of the air intake to the turbocharger by Cummins engineers.

There is some reason to believe that "clogged supercharger" may have been less than the full story of the failure of the Diesel to finish. The same cause was given in 1950, when the earlier car dropped out. In 1952 when it withdrew, the Cummins drove easily out of the pit area and back to its garage, leaving some to suspect that in return for allowing the big, oversize truck engine to run at Indianapolis, Cummins had promised it would not win the race.

THE KK500

The independent front suspension of the KK3000 led to heavy tire wear. Many examples were not well set up, due to most mechanics' lack of experience with the arrangement. Race "wrenches" were used to cars using Ford-style leaf springs dating to the Model T days. Torsion bars were simply lighter and more adjustable leaf springs to the racing fraternity, as long as they were attached to solid axles. Solid axles provided firmly set camber and allowed caster to be adjusted and locked in position. Independent suspension was subject to varying caster and camber and other difficult-to-control parameters.

Kurtis completed the design of his next Indianapolis car, the KK500, shortly after finishing the Cummins chassis. He still was not immediately able to sell either the offset or the laydown engine idea. Eugene Cassaroll ordered a KK500 for an Offy and Cal Connell one for a Cadillac engine, both for upright mounting. The Connell car in fact had independent front suspension. It was shipped, incomplete, to Detroit, where Ed Gdula finished it.

Seeing the problems with independent front suspension, Kurtis laid out his new car with solid axles, the front axle located by leading arms and torsion bars. The rear axle was still located by hairpin radius rods and parallel torsion bars. Side movement and roll centers were controlled by Watts linkages and Jacobs ladders. (The latter trace their lineage to old-fashioned telephone mounts.)

Howard Keck, of the Superior Oil Company, had two very sharp mechanics in Jim Travers and Frank Coon, whom Kurtis persuaded of the benefits of the laydown engine in the Cummins. On March 1, they ordered a car to be built with the 270-cubic-inch Offenhauser engine (Serial Number 130) offset 9 inches to the left and inclined 36 degrees to the right. At the time, this was believed with some reason to be as far as an Offy could be tilted without causing problems with the oiling system.

The KK500 frame was built of 4130 chrome-moly steel tubing with a hat-section steel plate welded vertically on each side, providing an external belt rather analogous to the stiffeners in the KK2000 chassis. The offset engine shifted about 50 pounds static weight to the left side, tending to counteract centrifugal force in the turns. The Keck steering, insisted upon and supplied by the team, was made by Manning, rather than the standard Casale unit.

Travers and Coon were gifted, careful mechanics and, with one tiny exception, the Keck car was beauti-

Frank Kurtis watches as Rudy Capricana applies final touches to the 1955 KK500C built for Firestone as a tire test car, using the big Chrysler Hemi V-8 engine. *Arlen Kurtis collection*

Firestone

Bill Vukovich's *Fuel Injection Special* Number 26, a KK500 roadster, is hauled away on the hook after his steering failed at Indianapolis in 1952, nine laps from the finish. *Author's collection*

The Chrysler team in 1952. Chrysler financed three Kurtis KK500As to carry its hemispherical head V-8 engines. The Number 17 car is the *Shouse Motors Special*, the Number 66 is the *Wolcott Chrysler Special,* and on the right, being started, is the second *Wolcott Chrysler Special*, Number 25. Although Joe James and George Connor had tested the Number 25 car at competitive speeds, the AAA withdrew its plan for a 366-cubic-inch stock block rule, and at 270 cubic inches the cars were too slow to qualify. *Author's collection*

fully prepared. They hired as a driver Bill Vukovich, a taciturn but talented midget and roadster pilot from Fresno, California, who had qualified once at Indy in three years of trying, but showed promise. Vukovich took a look at the Keck car and said it reminded him of the track roadsters he had driven in California, and that name stuck to the offset-engined cars.

On race day in 1952 Vukovich, starting from the middle of the third row, was one of those who passed Agabashian on the first lap, and he and young Troy Ruttman started a battle royal that went on for most of the race. Ruttman was driving J. C. Agajanian's Kuzma upright dirt car, similar to a KK2000. A total of 21 Kurtis chassis started the race, including Jim Rathmann and Sam Hanks on the front row and Nalon and Miller in the Novis. It was the first time that cars by a single builder had made up nearly two-thirds of the entire Indianapolis field.

Late in the race Ruttman began closing on Vukovich. With about 20 laps to go, Troy was 20 seconds behind and gaining a second on each circuit. As the laps reeled off, Vukovich suddenly was seen to be wrestling with the steering and finally on Lap 191, he slid into the third turn wall with little damage except to his pride and pocketbook. Ruttman took the race, the last to be won by an upright dirt car. For Vukovich, something had gone adrift in the Keck car's Manning steering. In the aftermath, there were rumors that Kurtis-Kraft had not properly tested it, while a hurt Frank Kurtis maintained he had objected to it from the beginning and had directed Travers and Coon to check it before delivering it to his shop.

Despite Vukovich's problem, the superiority of the KK500 design was immediately apparent, and after the 1952 race Kurtis received another order for one of

66
Johnny Pawl
RACING EQPT. Co.
RACE TEAM

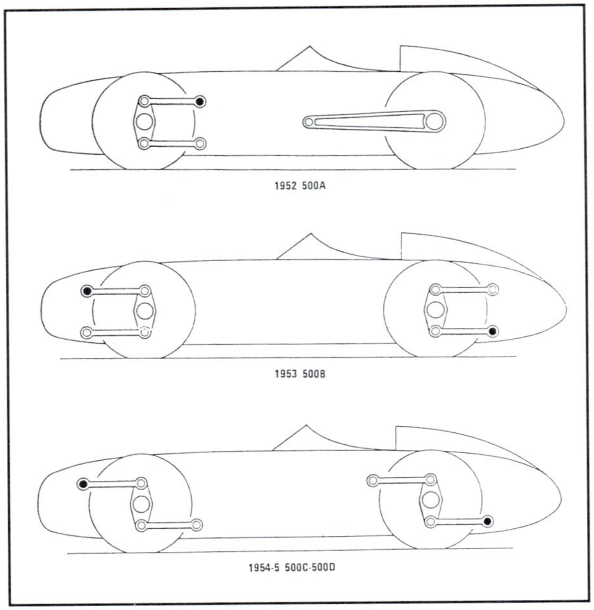

The original Kurtis Indianapolis roadsters of 1952–1955 were quite similar in their basic construction, with solid axles front and rear, but there were three variations in the torsion bar suspension systems: In 1952 the rear axle was supported on parallel torsion bars and located by hairpin radius rods. The front torsion bars were placed behind the axle, which was suspended by a pair of leading arm links. For 1953, on the KK500B design, the radius rods were eliminated, and the rear axle was suspended via transverse torsion bars with leading arms, while the front used leading arms from crossed torsion bars. In 1954 the KK500C and in 1955 the 500D had the torsion bars moved as far fore and aft as possible, with the idler linkages reversed from the 500B arrangement. The rears were located with Jacob's ladder linkages, while the front used a Watt's linkage. This layout set the style for all subsequent midget, sprint, and dirt Championship suspensions (with the exception of the coil-overs) for the next half-century in American dirt track racing. *Frank Kurtis*

the KK500s from Roger Walcott, to carry a big Chrysler Hemi engine. With Firestone picking up the tab for track rental at Indianapolis, standby ambulances and other costs, Walcott tested the Chrysler under a proposed 330-cubic-inch rule for stock block engines. The engine was said to produce 400 horsepower at 5,200 rpm on alcohol. Driven in October 1952, by Joe James and Joe Sostillio, the Chrysler-powered car covered more than 500 miles over three days at speeds equal to or in excess of the track record. Seeing that, and the 1952 pole run by the big diesel, the Indianapolis establishment, with its investment in 270-inch Offenhauser engines, immediately put the kibosh on special breaks for stock blocks. Cal Connell with his Cadillac; Walcott, with another KK500 chassis; and Belanger, a Chrysler dealer with his own KK500, tried to run the stock block V-8s in 1953 under the old 270-cubic-inch rule, but their cars were not quite powerful enough to qualify at that displacement.

Bill Vukovich in the Keck KK500 roadster, the 1953 Indianapolis winner. This was the first Indy car to be called a roadster, and it set the style of Championship cars for the rest of the front-engine era. Other KK500s were built in its first year, 1952, but only the Keck *Fuel Injection Special* had the engine offset, the signature of the roadster type. *IMS*

Nineteen fifty-two was also the year in which speeds led to tire wear that dictated two fuel stops. No one after that would try to go the 500 miles on just one visit to the pits. The result was smaller fuel tanks with lighter loads of gasoline or methanol. The following year, Firestone released its new wide tires for 7-inch rims, replacing the 6-inchers that had been around since the 1920s.

The 1952 KK500s were redesignated 500As when a slightly improved model built for 1953 was tabbed 500B. The 500B model had the front torsion bars crossed in front of the radiator ahead of the axle and can be distinguished visually by the absence of the hairpin rear radius rods in favor of leading arms from transverse-mounted torsion bars.

Eight KK500Bs were built for the 1953 race. Aside from Walcott's underpowered Chrysler-engined car, an unfinished car sold to Ed Gdula that never made it to the track, and Ed Walsh's *Bardahl Special* Number 24 that Cliff Griffith wrecked in practice, five healthy new Kurtis cars made the race. Agabashian qualified Andy Granatelli's Grancor car second, behind Vukovich in the Keck KK500, with Jack McGrath in Hinkle's KK4000 on the outside, making the Indy front row again all Kurtis.

Frank Kurtis and Ted Halibrand work on the Belond KK500B Indianapolis car of 1953. Johnnie Parsons drove this car, but the crank broke on his 86th lap and he finished 26th. *Arlen Kurtis collection*

The *Belond Miracle Power Special* Number 33 of 1955. This car, a Kurtis KK500B, was driven by Johnnie Parsons in 1953 and 1954. Then Quinn Epperly built this streamliner body for it, and Jim Rathmann brought it home 14th. *Mark Dees collection*

Johnnie Parsons in the *Belond Equa-Flow Exhaust Special* Number 15, a KK500B that refused to restart in the pits at Indianapolis on its 79th lap and was credited with 32nd place. *Bruce Craig*

Frank Kurtis with the 270-cubic-inch Offenhauser engine for the *Sumar Streamliner* Number 48 of 1955. At Indianapolis, driver Jimmy Daywalt was unhappy with the streamlined body over the wheels, and the side panels were removed. The car still qualified at 139.416 miles per hour and finished ninth. *Arlen Kurtis collection*

In 1953 the first seven cars to finish were all Kurtises, as Vukovich made a runaway of the race. The eighth-place finisher was a car Myron Stevens had started and Kurtis completed for H. A. Chapman. Oddly, except for Vukovich, the top seven were closely bunched in three KK4000s, a KK3000, and two 500Bs. Looking at that mix of three quite different chassis, one could wonder what design was the fastest way around the 2-1/2-mile Indianapolis oval. In total, 21 Kurtis cars were in the field that year, including KK2000, KK3000, KK4000, KK500A, and KK500B models.

THE KURTIS ROADSTERS COME OF AGE

The 1954 Kurtis model was another roadster, the KK500C. Most of them were built with engines offset or laid over from 18 to 36 degrees. Because in 1953 Sandy

DU
CAMSHAFT
NN
DETROIT

Previous page
Harry Dunn and the *Dunn Engineering Special* Number 89 of 1954. Dunn was a talented cam designer and a persistent Indy car owner. Ed Gdula was supposed to have completed this Kurtis in 1953, but did not get it ready in time. Dunn entered it at the Speedway from 1954 through 1957, with its best finish, by Pat Flaherty, a 10th place in 1955. *John Fedricks collection*

Belond's 500B had suffered breakage of several welds between the tubular frame elements and the vertical strengthening plate, the 500C was a completely tubular structure with enough cross braces to carry the stiffening loads. The 500C frame was an inch narrower than the 500B and had refinements such as Dzus hood fasteners instead of the old leather straps and a Jacob's ladder arrangement to locate the rear axle. Ten were built, plus a special version for Firestone, beefed up to carry a Chrysler Hemi V-8. Firestone needed a tire testing car, but did not want to buy the supposedly more expensive Offenhauser engine.

The KK500C cars were immediately successful at Indianapolis, as Jack McGrath put the Hinkle Number 2 on the pole. They were less effective on dirt, in part because their turning radius was only 25 degrees, about 10 degrees less than was believed needed on dirt tracks. They did have about 20 percent smaller frontal area than the KK4000s and a lower drag coefficient. Of the 33-car field, 21 were again Kurtis-built. In the race, Bill Vukovich was for the second time unbeatable in the Keck 500A. Jimmy Bryan powered an ill-handling Kuzma upright dirt car to second place, with McGrath third in a Belond KK500C.

Troy Ruttman brought Gene Cassaroll's Auto Shippers KK500A home fourth, with the KK500C cars of Mike Nazaruk and Freddie Agabashian fifth and sixth. The Kurtis roadsters had finally achieved their precarious reign at Indianapolis, taking five of the first six places.

1955

For 1955 Howard Keck planned a radical new streamliner for Vukovich to drive in place of his by-now three-year-old KK500. It would be very lightweight with much magnesium, and would be powered by a 270-cubic-inch V-8 engine based on a DeSoto block modified with dual overhead cams. Leo Goossen, the Meyer & Drake designer who had drawn the Novi V-8 and the Novi chassis, drafted the engine modifications on a sheaf of blueprints, but the project faltered. A test engine was never brought to race-ready shape. The light streamliner never raced.

Vukovich and his talented team of mechanics, Jim Travers and Frank Coon, were offered Lindsay Hopkins'

The *Braun Birch Special* Number 67 of 1957, a four-year-old KK500C that was originally the Lindsey Hopkins Number 35 that Pat O'Connor drove. This is the car that Bill Vukovich was driving in 1955 when he was killed. *Bob Tronolone*

Bob Sweikert, winner of the 1955 Indianapolis 500 in a Kurtis KK500D for which A. J. Watson was the chief mechanic. It was the last Indy victory for a Kurtis-built car, as Watson and his disciples took over the winner's circle for the next nine years. *IMS*

year-old Kurtis KK500C that had been driven to 21st place in 1954 by Pat O'Connor. (O'Connor spun out in the southeast corner on his 181st lap that year.) Travers and Coon went over the Hopkins car with meticulous care, and Vuky seemed to have lost none of his touch as he put the car into the middle of the second row with a 141.071-mile per hour qualifying run.

Despite the best efforts of designers and mechanics, things go wrong. Vukovich's steering broke in 1952 with little damage except to his pride and pocketbook, but in 1955 during practice at Indianapolis, mechanics on the Peter Schmidt team failed to sufficiently tighten a set screw holding the steering arm to the shaft on Manuel Ayulo's car, a KK500C. The arm came loose and Ayulo was killed. AAA officials immediately decreed that steering arms must be bolted, not simply secured by a set screw.

At that time the latest Kurtis model, the KK500D, was a very slightly refined version of the 1954 car, with

The *Sumar Streamliner* at Daytona International Speedway in 1959. It ran at Indianapolis in 1955 and 1956. Marshall Teague tried to set a closed-course speed record with it in 1959 at Daytona, but was killed in the attempt. *Author's collection*

Art Cross makes a pit stop in 1955 in his Belanger KK500D roadster. The car finished 17th when a rod broke in the Offenhauser engine. *Bob McConnell collection*

provision for power steering, although it was never used. A. J. Watson bought one, unfinished, for John Zink's driver, Bob Sweikert.

There were 20 other Kurtis chassis on the track as the field pulled away from the starting line. Ten of the first 12 starters were Kurtis creations. As the cars began to stretch out, Bill Vukovich, as expected, charged up from the second row to take the lead. It began to look like a third straight runaway for Vuky, but it was not to be. On his 56th lap, coming out of the southeast turn, he encountered a three-car tangle. Before he could dodge the melee, another car rammed his and he went end over end, out of the track to his death.

Three hours later, a subdued Indianapolis crowd watched Sweikert in Zink's KK500D take the checkered flag and $76,000 in prize money. Kurtis chassis swept 8 of the top 10 spots, but it was not the same as in 1954, when Vukovich had been a grand two-time winner. Vuky was gone and there would not be another like the quiet Slav from Fresno. Jimmy Bryan and then Rodger Ward would inherit the mantle, but not until a young rookie then still driving midgets and sprint cars, A. J. Foyt, appeared, would Indianapolis see the likes of a another Vukovich.

Frenchy Sirois and Tiny Worley prepare the *Belanger Motors Special* Number 99, a KK500D that Art Cross drove at Indianapolis in 1955. *Bob McConnell collection*

These are the Novi-powered cars that Kurtis built in 1956 as his KK500F models. They suffered the common fate of the Novis, with their best finish Paul Russo's 4th place in 1957. *Arlen Kurtis collection*

With a great finish in 1955, despite Vukovich's death, Kurtis looked forward to 1956 with anticipation. He had sold only six KK500Ds in 1955, but seemed to be doing well with projects for the aviation industry, the sports car crowd, and the custom car market. His latest creation, the KK500E for Dan Levine's Federal Automotive, was a 500D with the Offenhauser engine tilted to the left. He was building two new unique rear-drive KK500F cars for Lewis Welch's Novi engines that kept the shop busy and still in the thick of racing design.

The new Novis were the rear-drive cars that Kurtis had been advocating to Welch for six years. Built as roadsters, their Novi V-8 engines could not be offset to the left quite as far as the narrower Offenhauser engine allowed, but the 82-gallon fuel tanks and other weights were put on the left to maximize weight on that side. The new cars were given a massive tail fin that was actually more cosmetic than aerodynamic.

But Kurtis' business suddenly went sour in late 1955. He had gotten out of the partnership with Ed

Shorty Templeman in the Arcerio Maserati, the same Kurtis KK500C that Bill Vukovich was driving in 1955 when he was killed. Lindsey Hopkins repaired the car after Vukovich's death, and Jim Rathmann finished 20th in it in 1956. Doug Carruthers entered it in 1957 but could not get it qualified as the Braun Birch Number 67. Tony Parravano bought the car a year later to carry the Arcerio Maserati, but Shorty could not wring enough speed out of it. *IMS*

Mike Nazaruk, an ex-Marine, in the McNamara Number 73, a KK500C roadster, in 1954. Mike came in fifth at Indianapolis in this car for owner Lee Elkins. *Tom Saal collection*

George Bignotti and the Bowes crew service Johnny Boyd's Kurtis KK500G2 at Indianapolis in 1957, when Boyd finished sixth. *Author's collection*

Walsh by joining in a corporation that included Walsh, Robert McCullough, Buick dealer Bill Murphy, Ford dealer Jim Coberly, and Jack Hinkle. The corporation went bankrupt, and Kurtis was forced out of it just as he was finishing six new cars to run at Indianapolis in 1956. He completed the two new rear-drive Novis for Lew Welch at Alger Street in Los Angeles, but he lost many of his tools, most of his equipment, and all of his records. He had to start over at a new place, at 525 East Colorado Boulevard in Glendale.

Twenty-three Kurtis chassis would take the green flag in 1956, all but one the 500 models built since 1952. The exception was Sam Traylor's outdated KK4000, driven by Al Keller, a journeyman Eastern midget driver who proved to be a talented pilot at Indianapolis. After Keller's 14th place finish, a dirt track car would never again qualify for the 500.

George Bignotti, a bright young mechanic, and George Walther Jr. had ordered the new KK500G models, slimmed-down versions of the roadsters with the torsion bars mounted under the axles. The Bignotti car was the lightest Kurtis yet, with many parts made of

Bill Boyd heads for the garage at the end of the 1958 500 in the Bowes KK500G2 Number 9, in which he finished third. *Tom Saal collection*

Opposite page
George Amick in the *Federal Engineering Special* Number 4 a KK500D, at Indianapolis in 1957, when the car did not qualify. Freddie Agabashian had driven the car in 1955 and 1956, with its best finish a 12th in 1956. *Bignoni collection*

costly titanium. It finished 15th, with a blown engine. Only four Kurtis products finished in the first 10 at Indianapolis in 1956, despite an overwhelming number of Kurtis' cars in the starting field. An aggressive young mechanic by the name of A. J. Watson emerged as both the builder and the mechanic of the winning car.

In 1956, Pat Flaherty, in his Watson, would edge Sam Hanks' Kurtis by 0.187 miles per hour, and never again would a Kurtis see the checkered flag first at Indy.

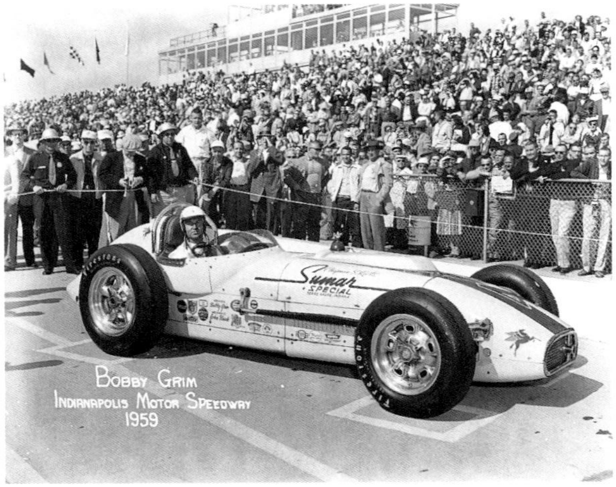

Bobby Grim in the *Sumar Special* Number 48, a KK500G2 chassis. Grim finished the 500 in 26th place in 1959, the car's fourth year at the Speedway. Pat O'Connor put this car on the pole in 1957 and finished 8th that year. In 1958, it was involved in the huge first lap crash, and O'Connor was killed. *Tom Saal collection*

Watson's *John Zink Special* was an admitted copy of the 1955 Kurtis, but slimmed down and lightened with magnesium in place of aluminum and with as many other weight-saving shortcuts as possible. Kurtis refused to use magnesium, because he felt it was dangerous to work with and a fire hazard in a race car, but it did help Watson achieve a weight saving of 200 pounds. Watson also had engineering help from LeRoy Payne and Hank Plum in designing the truss members of the frame, placed by Kurtis by eye but by Watson in a more sophisticated manner.

The problems with his business partners made it hard for Kurtis to concentrate on designing and building race cars. Contributing to the difficulty was Kurtis' practice of designing his new cars according to his idea of what should work on the track, rather than by asking drivers and mechanics for their opinions and observations. Years later he would say, "Except for Clint Brawner, not one man, not an owner or a mechanic, came to me with an idea." Another exception to what Kurtis felt was lack of appreciation of his innovations was George Bignotti, who succeeded in having Kurtis build three lightweight cars for the Bowes Seal-Fast team in 1957.

In Kurtis' opinion, most mechanics were inclined to do things the way they had always been done, or to have ideas that were unworkable. Barry Knowlton, one of his employees, said years later that Kurtis, except for Indianapolis, had quit going to the tracks. In an interview Frank tape-recorded in 1982, Kurtis himself confirmed that he built his cars largely to his own ideas and his alone.

"A customer never had a single thing to do with design of my cars. It was left entirely up to me," he said years later. The mechanic on the 1953 Auto Shippers car objected to offsetting the engine, and it was mounted in the center, and Pete Clark and Rex Mays insisted on a rail frame chassis, which Kurtis felt was a poor design for the 1947 Bowes car. It was frustrating to Kurtis when his latest designs were not accepted.

Gathering his team of employees together again after the McCullough debacle, Kurtis resumed work at the new shop on Colorado Boulevard. There is some slight uncertainty about the nomenclature of the 1957 production, which surviving Kurtis records show as all G2 models. The G and G2 models had the torsion bars mounted in front of and below the front axle. Despite the fact that Pat O'Connor put his Sumar Number 12

Federal Engineering
DETROIT SPL.
BEAR
Firestone
95

KK500G on the pole in 1957 at 143.948 miles per hour and three 500Gs finished in the top eight, the design was not particularly successful. It did not win the 500. At the time Eddie Kuzma had also built several cars with low-mounted torsion bars and they were not successful either, leading to a feeling at Indianapolis that the design was at fault.

Juan Fangio, the Argentinean world champion, took his rookie test at Indianapolis in the *Dayton Steel Special* Number 77, a KK500G, in 1958. Although he passed his test, he decided not to race in the 500. Mike Magill qualified the car and finished 17th. *Bruce McCaw collection*

The 1957 500 was the year that George Salih's laydown, the first car to have the engine laid over to 72 degrees, was the magic ticket. Kurtis said later that the Salih car, built by Quinn Epperly, was, aside from the placement of the engine, a complete copy of the Kurtis KK500G of 1956.

"Salih worked for Meyer & Drake, who built the Offenhauser engine," Kurtis noted. "He was able to work out the changes in the oiling system that allowed him to put the engine on its side. We did not have that capability." Salih and driver Sam Hanks financed the car themselves, and it proved to be a winner. Hanks qualified it 13th, but with less tire wear and excellent handling it won the race.

The record shows that two KK500Gs went out of the 1957 500 with leaking fuel tanks. The same malady struck the Kurtis entries at Monza, Italy, later in 1957, where the speeds they encountered on the rough concrete

Paul Goldsmith in Smokey Yunick's *City of Daytona Beach Special* Number 31 at Indianapolis in 1958. The car, a KK500G2, crashed on the first lap. *Armin Krueger*

track beat the cars' tanks apart. The fuel tank leaks hurt Kurtis' reputation at a critical point in his career. Only two new Kurtis cars were ordered for 1958, one of them a tire test vehicle for Firestone. Roy Sherman had Kurtis build a car tabbed as a KK500H with an unusual independent front suspension.

In 1958 both Watson and Kurtis experimented with offsetting the entire bodies of their cars to the left by between 1 and 2 inches. Watson's innovations included onboard air jacks to speed up tire changes during the race.

At Indianapolis that year there were only 16 Kurtis-Krafts in the race, 6 fewer than in 1957. Five Kurtises, including the Bowes and Sumar cars, were wrecked in the first-lap carnage in which Pat O'Connor was killed. Boyd, in the Bowes KK500G, finished third as the Salih laydown won for the second year in a row, this time with Jimmy Bryan at the wheel. The Sherman D-A Lubricants Kurtis lost its brakes and damaged its steering on the 52nd lap. At the second (and last) Race of Two Worlds at Monza, 8 more Kurtis chassis suffered fuel tank leaks. The result was another dry year for race car orders for Kurtis, with only George Bignotti and Roy McKay stepping up to buy 1959 models, then called KK500J.

In 1959 there were 30 Kurtis entries at the Speedway but only 12 qualified. Two were wrecked in practice and several were out of date chassis, no longer competitive. A Kurtis chassis was no longer an almost automatic ticket to the race. Duane Carter in Smokey Yunick's *Reverse Torque Special* was, in seventh place, the best Kurtis finisher. Four Kurtis drivers, Chuck Weyant, Jud Larson, Mike Magill, and Red Amick, went out in one third-turn crash on Lap 45.

In 1960 only William Tucker, of Galion, Ohio, stepped up to order a Kurtis Indianapolis car for his Tucker Electric Company sponsorship, a KK500J2. The J2 was a modest upgrade from the 500J with the torsion bars placed high in front and with the engine offset, but in Watson fashion, upright. Veteran driver Cliff Griffith failed to find enough speed to qualify Tucker's Kurtis.

There were 24 Kurtis entries in 1960. Four were wrecked in practice laps by less than skilled drivers, and 12 were either too slow or were unable to make qualifying time. Kurtis himself was happy not to be commuting to Indianapolis in May, and Frank Kurtis Inc. was busy with industrial products and work for the aerospace companies in Southern California. Eight Kurtises made the race. Bob Christie was 10th in the first Kurtis to finish, the Federal Engineering KK500D, built five years earlier.

The Sumar crew works over Johnnie Parson's *Sumar Special* Number 18 in 1957. A Kurtis KK500G, it finished 16th. The car never again qualified in four years of trying. After Chapman Root, subsequent owners were LeRoy Foutch Jr. and Harry Beck. *Bob Tronolone*

Nineteen sixty-one was a replay of 1960. Kurtis built a KK500J2 with a roll bar for Fred Gerhardt and another with built-in air jacks for pit stops for New Yorker Anastasis Vatis. There were 15 Kurtis entries that actually reached the Speedway and 4 qualified. Chuck Hulse wrecked the Vatis car in practice. Jack Turner got the Gerhardt car into the race but on his 25th lap, Don Davis in Floyd Trevis' *Dart Kart* Number 83 blew an engine on the main stretch and spun in his own oil, collecting Turner's Kurtis which went end over end, fortunately without serious injury to the driver.

Fred Gerhardt's *Las Vegas Club Special* Number 45, a Kurtis KK500G2 that Bill Cheesbourg was unable to qualify in 1957. *Bob Tronolone*

By 1962 there were 16 Kurtis entries, of which 5 went to the starting line. Frank built two new rear-drive chassis for Andy Granatelli, then the new owner of the Novi supercharged V-8 engines, but Granatelli was no luckier with the powerful Novis than Lew Welch had been. Neither car qualified. Jim McElreath was the best Kurtis entry, finishing sixth in Ollie Prather's six-year-old KK500G2, which had started out as Ray Crawford's *Mirror Glaze Special* in 1957.

Frank Kurtis still had a touch with the setup of his cars, if anyone would listen. Bob Christie, driver of William Tucker's North Electric Number 29, a two-year-old KK500J2, recalled years later that they were having trouble before the 1962 race going as fast as they had in 1961. The car was simply not handling. Christie said the front end was hopping all over the place. Kurtis came over and asked if they needed help. "That old man told us to reduce the camber in the front end and how to set the rear bars, and we went four miles an hour faster," Christie said.

Art Malone's *Mad Dog III*, a Chrysler-powered, blown, former Firestone test car, a special Kurtis chassis that broke the 180–mile per hour "barrier" at Daytona International Speedway in 1961. Bob Osiecki installed the engine and a pair of inverted airfoils to keep the car on the ground and avoid its becoming airborne like Marshall Teague's *Sumar Streamliner*, two years earlier. *Art Malone collection*

The *Bowes Special* Number 14 of Freddie Agabashian, a Kurtis KK500G2. This car finished 22nd in 1957 after the fuel tank split. *Bob Tronolone*

Bob Boyd's *Bowes Seal-Fast Special* Number 6, a 1957 Kurtis KK500G2. Boyd brought the car in 6th at Indianapolis that year. *Bob Tronolone*

Jim Hurtubise in the Novi is pushed out on to pit road at Indy for a test run on Carburation Day in 1962. Neither Novi had qualified, but Chief Steward Harlan Fengler allowed Andy Granatelli to make a test run to see if various problems had been solved. *Bob Tronolone*

In the race Christie, Chuck Rodee, Allan Crowe, and Jack Turner tangled on the front straight, putting all four out of the 500.

In 1963, of the six Kurtis entries only the three Novis qualified: the two cars built in 1962 and Art Malone's 1956 Lew Welch chassis, which finished 31st.

The final Kurtis Indianapolis chassis (Serial Number KK500LC-63-126) was built for LaPorte, Indiana, plastics manufacturer John Chalik. Chalik let it be known that his racing background included service as a riding mechanic in a 1919 board track race at Cincinnati, but he seemed receptive to new ideas. He had entered a car at Indianapolis in 1961, a Trevis chassis powered by a 255 Offy. Gene Hartley had driven it to 11th place and it was running at the end, two laps down, not too bad for a first-year team. However, another Chalik car had lost its magneto during the race.

Chalik's mechanics, Joe Scopa and Barney Christensen, approached Kurtis with a plan to put NASCAR ace Junior Johnson in an Indianapolis car. Johnson was a tough southern redneck who had served time for bootlegging. He wanted a try at Indianapolis, which at the time was still the

Andy Granatelli and Bobby Unser, left, look at a problem with the supercharger on the Studebaker STP Special #6 in the Indianapolis pits in 1963. Unser hit the wall on lap two of the race that year.
George Peters collection

Jim "Herk" Hurtubise in one of Andy Granatelli's Kurtis-built Novis, a KK500K, in 1963. Hurtubise was stopped by an oil leak and finished 22nd.
George Peters collection

pinnacle of American racing. Johnson visited Glendale to be fitted to the car. He observed that he was not entirely comfortable with open cockpit cars, having driven all his racing career in coupes. In response, Kurtis laid out a fairly graceful roll cage for the car, to give Johnson a bit more security.

The Chalik car incorporated fully independent suspension and, at Chalik's insistence, dual battery-powered ignition. Having had a magneto failure on an earlier car, Chalik was determined to avoid the problem in 1963. The Champion Spark Plug Company tested the Chalik system on the dynamometer and found it 75 horsepower below the output of the same engine with a magneto. Chalik, however, was adamant that the battery ignition stay.

Dick Rathmann test-hopped the car as soon as the Speedway opened and told Kurtis it handled well. "Never had a car that steered so good," Rathmann gushed.

Johnson began the three-stage driving test at Indianapolis, as a new driver running at graduated speeds until veteran observers were satisfied that he was competent at racing velocity. Johnson handled the first phase, but in the second, the open cockpit bothered him. Kurtis said, "He told me it felt as though he was in the

Jim Hurtubise in the Novi getting ready at Indy for a test run on Carburation Day at Indianapolis in 1962. Neither Novi had qualified, but Chief Steward Harlan Fengler allowed Andy Granatelli to make a test run to see if various problems had been solved. *Bob Tronolone*

nose cone of a missile." Johnson added, "Indy is not for me," and went home to Georgia. Kurtis took off that same day to return to Glendale.

A week later, Chalik called Kurtis and begged him to come back to Indianapolis for time trials. When he arrived in Gasoline Alley, Frank found chaos in the Chalik team. Barney Christensen was having trouble getting the other Chalik car, an older Trevis, up to speed. Bob Harkey eventually crashed it in practice. Having fired Joe

Art Malone in the *Studebaker STP Novi Special* Number 3 in 1964, when the ex–drag racer finished 11th at Indianapolis. Granatelli hired Malone, who had no Indianapolis experience, and bought a used Kurtis dirt car for him to drive for enough oval track experience to be allowed to drive at the Speedway. *Art Malone collection*

The first eight Championship cars Frank Kurtis built were individually unique and only the *Kurtis-Kraft Special* Number 7 was built to his own design. The 1952 *Cummins Diesel Special* Number 28 was also unique, designed and built around a 401-cubic-inch Cummins truck diesel engine. The Pierson cars were built to a front-wheel-drive design by Leo Goossen. Three of the big cars were actually dirt sprint cars. One was built to a special design for Firestone for tire testing. There remain 100 Kurtis Championship cars that Frank Kurtis manufactured, to 16 different designs.

The *Kurtis-Kraft Special* of 1948 was also the KK1000, though it was not widely called by that name, and only one, the Parsons 1950 Indianapolis winner, was built to that design.

Johnnie Parsons in Ed Walsh's *Wynns Special*, a KK4000, Serial Number 342, at Littleton, Colorado, in 1951. Parsons finished 14th after he lost his brakes. *LeRoy Byers*

The Jones & Maley Number 95, a KK500C, was driven to 13th place at Indianapolis in 1957 by Bob Christie. *Bob Tronolone*

KK2000

The first Championship car that Kurtis built in quantity was the KK2000, his version of the standard upright, solid-axle American single seater, laid out by Harry A. Miller in 1923. It was an enlarged, 96-inch-wheelbase copy of the tube-frame, torsion bar Kurtis midget that had swept all before it in its class on dirt and pavement for nearly three years. The KK2000 has two parallel tube rails, joined by a perforated stiffening plate. The mounts for the hairpin shaped radius rods are semicircular and easy to spot. As far as is known, only George Tuffanelli's race manager, Charlie Pritchard, changed those mounts. Hal Cole and Charlie Engle had shortened KK2000s built as sprint cars. The only difference between the champ and sprint cars is their length. The sprint cars have a wheelbase of 84 inches. Approximately 13 KK2000s were built between 1948 and 1950.

Number 99

The Meyer & Drake, later the Bellanger, Number 99 was essentially a Kurtis midget stretched out by a little more than 2 feet. The front torsion bars were mounted above the axle and outside of the body. Two were built. The second, for Johnny McDowell, was cut up after he was killed in its second race.

KK3000

The next design that Kurtis built was patterned after the successful *Kurtis-Kraft Special* of 1948. It had the same independent front torsion bar suspension using 1949 Ford parts and a solid Halibrand rear axle, also suspended by torsion bars. The frame was welded of chrome-moly aircraft tubing. The body was wider and much more rounded than the KK2000, with a wide shovel nose. The rear axles were located by the usual hairpin radius rods. About eight KK3000s were built in 1950 and 1951. They were far less successful on the dirt tracks than had been the prototype *Special*.

KK4000

The KK4000 was a throwback to the KK2000: an upright car with solid axles and radius rods fore and aft. The frame was part tubular and part formed of perforated plate. Suspension at both front and rear was by torsion bars. The KK4000 was designed primarily as a dirt track car that could also run at Indianapolis. Though heavy for the dirt tracks, it was stout and reliable. These cars were hardly distinguishable from the dirt track cars made by Eddie Kuzma, Jud Phillips, A. J. Watson, Wally Meskowski, and other builders of the day who, in truth, had themselves copied the basic layout of the Kurtis midget. At least 15 were built between 1951 and 1954. Though not run at Indianapolis after 1956, they competed on the dirt until at least 1970.

KK500 and KK500A

The KK500 was the outgrowth of the design Kurtis came up with for the Cummins Diesel Engine Company in the fall of 1951, laying the engine over 85 degrees to lower wind resistance, and offsetting the weight to the left side. Kurtis had learned a great deal from the wind tunnel testing of the Cummins car. Howard Keck and his team of Travers and Coon were persuaded that offsetting the engine weight would improve handling and reduce tire wear. The KK500 was a wider, lower, version of the *Kurtis-Kraft Special*, with the engine set to the left, allowing the driver to sit alongside, not over, the driveshaft. The KK500 had a tube frame stiffened by a horizontal plate that

The pace lap for the 1958 500, when the front row ran off from the field on the parade lap. The second row, shown here, consists of Bob Veith, in the Number 14 Bowes car, a KK500G2; Pat O'Connor in the Sumar Number 4, another KK500G2; and Johnnie Parsons in the Number 45 Gerhardt, a KK500G. Veith and O'Connor were involved in the first-lap crash and O'Connor was killed. *Armin Krueger*

Roy Sherman with his special KK500H, built in 1958. This car had a unique front suspension that was not a success. In 1959 Smokey Yunick put his reverse-torque Offenhauser engine in the car and Duane Carter drove it to seventh place. *Edy Kurtis collection*

formed part of the body skin; solid axles, with typical Kurtis torsion bars and radius rods at the rear; and torsion bars in the front, behind the axle, which was suspended by leading arms. The grille teeth were formed of tubing, and were chrome-plated. Only the Keck KK500 was a true roadster with the engine offset. Four others were built, designated KK500A, but three of them had stock-block V-8 engines that were impractical to offset, and the *Auto Shippers Special* of Gene Cassaroll had its Offy tilted but not offset. The V-8–powered Belanger car had a concave grille, while the others had convex bars, vertical on the Walcott and Keck and horizontal on the Auto Shippers.

KK500B

The B model of the KK500 eliminated the rear radius rods for leading torsion bar arms, watts linkages, and Jacob's ladders to locate the axle. At the front, the torsion bars were moved ahead of the axle. Later versions of the KK500B had elongated tails and recessed grilles, but otherwise were basically little different from the tube frame solid-axle A models. All of the Offy-engined versions had the drivelines offset to the left, and two had engines tilted by 19 degrees. Eight were built, all before the 1953 race.

KK500C

The C model relocated the front suspension pieces, with the lower front linkage now located behind the axle. The radius rod attachment on the left side of the front axle was left free to rotate, while on the right it was solidly welded in place. Engines were almost all offset to the left, though some exhausted out of the left side and some out of the right. Most of the grilles were now made of chrome-plated mesh to keep foreign objects out of the radiator. A total of 11 500Cs were built, all in 1954.

KK500D

The KK500D was built in 1955. It was a slightly refined KK500C with no easily recognizable differences from the KK500C. The visible differences that did exist were largely minor, installed by each team on its own. The D Model did, however, have its fuel filler cap on the left side of the headrest. Seven were built, including the Sumar car, which was given a special streamliner body.

KK500E

Only one KK500E was built, for Federal Engineering, in 1956. It has the fuel filler on the left side of the upper tail surface, not in the headrest. On the right side the steering used an external drag link. The engine was tilted to the left rather than to the right, as was usual.

KK500F

The two KK500F cars were built in 1956 for Lewis Welch, as rear-drives to carry the ill-starred Novi engines. They had leading/trailing link torsion bar suspension and prominent tail fins that were modified somewhat over the years that they ran.

KK500G

Only two KK500G chassis were built, both in 1956, one for George Bignotti and the other for George Walther Jr. They were lower than the previous KK500 models with drag link steering on the right and prominent hood scoops. Bignotti's had the exhaust on the left, Walther's on the right. The front torsion bars were placed below the axle. Those cars had an oval grille opening, surrounded by a chrome tube frame "bumper."

KK500G2

Nineteen fifty-seven was a year of turmoil for Kurtis' business. The KK500G2 model was only a modestly refined KK500G. Engines were placed on the left but the degree of tilt, if any, and intake and exhaust arrangements varied. Most had drag link steering, and the chrome-surrounded grille is the chief distinguishing feature. Fourteen were built, all in 1957.

A Car-Spotter's Guide to Kurtis Indianapolis Racing Cars (continued)

KK500H

In 1958 Roy Sherman of Racing Associates ordered this one-of-a-kind car as the *D-A Lubricants Special*. It had a tail fin, a horizontal knife-edge nose, and an odd, massive trailing-link independent front suspension that was not successful. Smokey Yunick chopped off the front end and installed a solid axle in 1958.

KK500J

Kurtis built two of these in 1959, for Bignotti and Roy McKay. The McKay car was a slightly updated KK500G with a conventional nose; the Bignotti car had a knife-edge front like the KK500H.

KK500J2

These cars had the torsion bars back above the front axle. They had upright, offset engines with extended, flattened noses. Steering was external drag link on the right side, and many had the then-common external oil tank on the left.

KK500K

These were two rear-drive cars for Andy Granatelli, who bought the Novi team from Welch in 1962. They were essentially KK500J2s adapted to carry the bigger Novi engine.

KK500L

The last gasp of Kurtis-Kraft as a race car constructor, the KK500L was built with an innovative independent front suspension and a roll cage, but because it never had competitive power at Indianapolis, it was never properly tested. Only one example was built.

Year by year at Indianapolis, perhaps even race by race, cars were changed. Headrests were added, cars were narrowed, fuel fillers were relocated, and other visible changes made. The same car may have appeared in a number of different forms after it left Kurtis-Kraft.

This KK500L in 1963 was Kurtis' last Indianapolis car. Built for John Chalik, it had independent front suspension and a roll cage for driver Junior Johnson. *Robert McConnell collection*

This rear view of the KK500L Kurtis built for John Chalik in 1963 reveals the rookie stripes for Junior Johnson, a stock car champion tapped to make his first attempt at Indianapolis. Despite the roll cage, Johnson decided against running at Indy and went home to Georgia. Chalik refused to run a magneto, preferring an inadequate battery ignition system. Other drivers were unable to get the car up to qualifying speed. *Robert McConnell collection*

A NOTE ON SERIAL NUMBERS

After building seven post–World War II Indianapolis cars, Kurtis-Kraft Inc. began to number it's chassis. The midgets, the sports cars, and the subsequent Indianapolis cars each had a small brass tag attached to the dash with the model, year, and serial number stamped on it. If records of those numbers and the buyers had survived, the historian's job would be much easier.

Unfortunately, when the McCullough-Kurtis partnership collapsed, most of the records disappeared. Historians have searched for them for 30 years without success. Only fragments of the build-list have come to light. A typewritten list that appears to be fairly accurate exists for the Championship cars built through 1956. A handwritten list for the years 1957–1965 was found in Arlen Kurtis' files. The numbers correlate well with those found on surviving cars, so that the Championship car list appears to be valid, though at least four cars were built that do not show up on it.

The entire list of upright midgets built between 1945 and 1953 has apparently been lost. Kurtis tried to reconstruct from memory the numbers of the first 10 postwar midgets, and while he could name some of them, his memory was apparently not perfect.

Kurtis himself said that the first 10 midgets were finished in the winter and spring of 1946 before Indianapolis. Kurtis-Kraft ran an advertisement in *National Speed Sport News* on February 28, 1946, with a photo showing six cars under construction and listing "cars delivered to or on order by" Roy Russing, Ed Haddad, Rex Mays, Lloyd Axel, Perry Grimm, Billy Vukovich, Gib Lilly, and Ed Wallace. The ad did not describe the list as in chronological order. In the same issue of the paper an *NSSN* columnist, John J. Gliebe, mentioned visiting Kurtis' shop and seeing cars being built for Russing, John Balch, and Wallace. Historians believe that the "Haddad" car was ordered by Balch, and the Vukovich car by Fred Gerhardt. Buzz Lowe ordered one of the first cars, but did not take delivery of it immediately.

Although most Kurtis midgets were numbered, anecdotal evidence suggests the first few cars built, before George Shilala became shop foreman, may not have been given serial numbers.

Beyond the first midgets, no list exists. Some later owners have been able to trace the history of their cars, but the trail, a half-century later, is a very cold one. The author traced his car, serial number 0-267-48, to the first owner, Roy Hagedorn, but for most of the hundreds of Kurtis midgets, there is no record of the original purchaser. The numbers consist of "O" for Offenhauser, "F" for Ford V8-60, and "D" for Drake, the three engines commonly installed, while the last two digits indicate the year of construction, 1946–1952.

There is no certainty as to the number built, though the highest serial number ever found is O-367-52, on a car Ken Hickey bought in November 1952, shortly before Kurtis sold the midget business to Johnny Pawl. (Pawl did not number the midgets he built.) A number 888 has been found, but it appears to be an error. Other cars, probably built from kits, exist, with number plates bearing an "X." Analysis of the "X" plates suggests that there were as many as 152 kit cars given numbers. Together with the possibly unnumbered first 10 cars, the total of midgets built was probably 532.

Frank Kurtis said in a February 1952 letter to *Speed Age* magazine that he had built 550 complete midgets, not including kits and parts. Allowing for some mental rounding-up by Kurtis, that total is fairly close to the estimates that the author and historian Bill Montgomery have made.

There were 10 roadster midgets built by Kurtis in 1958–1959 and certainly other midgets built from parts sold by Kurtis-Kraft. As many as three incomplete cars went to Pawl, along with the jigs and fixtures he bought in 1953. In the author's opinion, the figure 550 is as close as history is likely to get to the true number of midget racing cars that poured out of Kurtis-Kraft between 1945 and 1959.

far left
Stock car champion Junior Johnson tried out the Chalik Kurtis in 1963, but after taking part of his driver's test, decided Indianapolis was not for him. *Bob McConnell collection*

Arnie Knepper in Tassi Vatis' *Konstant Hot Special* Number 18, a KK500J2, at Indianapolis in 1965. This was the last Kurtis Championship car to compete in the 500. *Jim Haag collection*

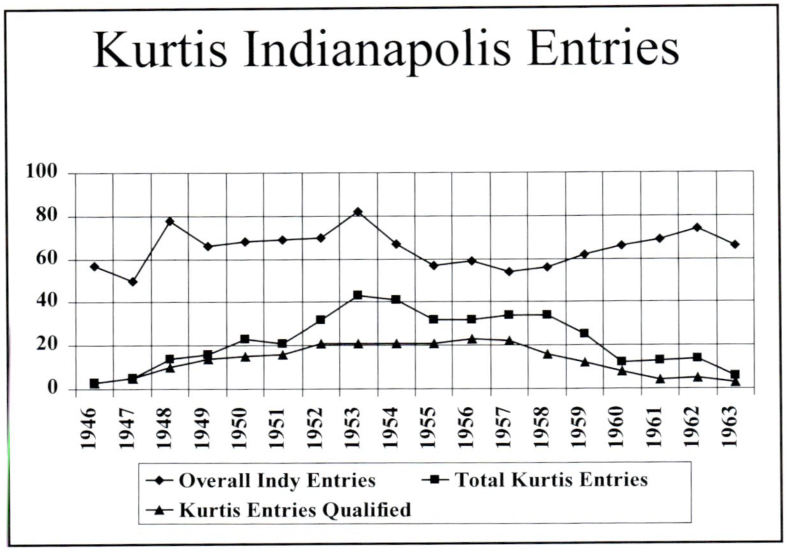

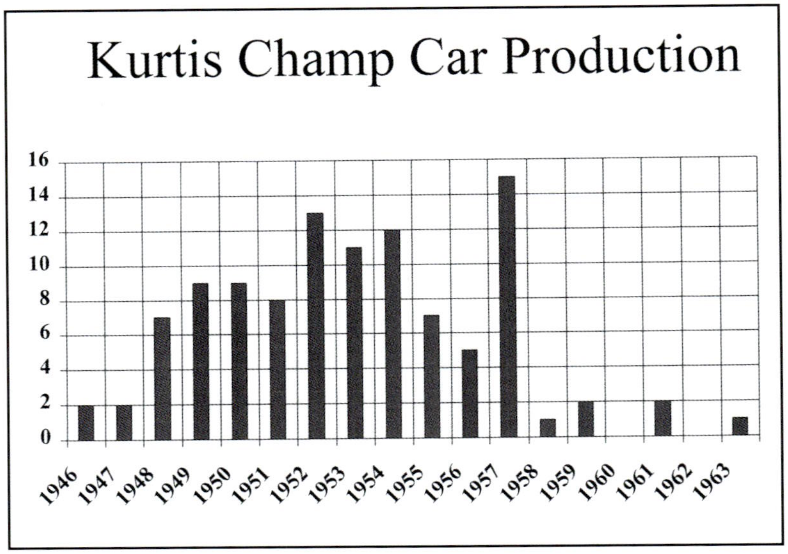

This is Tassi Vatis' number 18 car, a Kurtis KK500J2 at Indianapolis in 1965, when it finished 18th, the last Kurtis to race in the 500.

top left
Kurtis Indianapolis entries. Overall entries at the Speedway varied from fewer than 50 in 1947 to a high of 82 in 1953, probably affected in part by the national economy and the availability of sponsorship money. Entries of Kurtis-built cars climbed from 3 in 1946 to 44 in 1954, as older cars continued to be entered in successive years. The decline in 1955 to 1957 reflected both an overall decline in entries at Indianapolis and Kurtis' own unhappiness with the Speedway. Obviously, as only 33 cars can qualify, Kurtis reached a plateau in the race itself at about 21 cars from 1952 until 1957. Kurtis entries fell off rapidly after 1958, even while total Indianapolis entries climbed. As the front-engine era approached the end of its half-century reign, Kurtis was no longer represented there.
Angela White

top right
Kurtis Championship car production, 1946–1963. As Kurtis' cars became more successful at Indianapolis in the early 1950s, demand for them grew, then sagged after 1954, as Kurtis-Kraft became more involved with McCullough and the sports cars. In 1957, after he broke away from the McCullough operation, buyers returned, and his output of 14 KK500G2 cars in 1957 was his best year. However, none of those cars won the 500 and Frank became disillusioned with Indianapolis; the figures for 1958–1963 show that clearly.
Angela White

Scopa, Chalik at the last moment hired Bill Ward as crew chief on the new car and put Colby Scroggins in as the driver. Scroggins was a hard-bitten veteran of the West Coast roadster tracks but a rookie at Indianapolis.

Scroggins was able to get the Chalik Kurtis through the turns in good shape but the battery-powered ignition was cutting out at higher revs on the straightaways. Scroggins, Kurtis, Ward, and Christensen all begged Chalik to change to magneto ignition, but the owner refused. With qualifications about to start Kurtis said, "I looked at the mess Chalik had made of my car, and at Andy Granatelli, who still owed me money for the 1962 Novi chassis, and said the hell with it. I was sick. I climbed on an airliner and went home."

Kurtis cars would run at Indianapolis in 1964, and in 1965 Tassi Vatis' *Konstant-Hot Special* Number 18 would carry Arnie Knepper to 18th place. Scotsman Jim Clark won that 1965 Indianapolis 500 in a rear-engine Lotus. The day of the front-engine car, the roadster, and Frank Kurtis' Indianapolis career, were over.

The end for many Indianapolis roadsters in the 1950s was as a super modified running in a circuit centered on Oswego, New York. This example is the McNamara KK500C that Mike Nazaruk drove for Lee Elkins to fifth place in 1954. Jim Muldoon drove it years later at Oswego. *Bob McConnell collection*

Gordon Herring at Devil's Playground on the way up Pike's Peak in 1964 in the former Jack Hinkle Kurtis KK4000 championship car, then sponsored by Cummins of Colorado. *LeRoy Byers*

110

Kurtis Sports Cars

CHAPTER 5

Frank Kurtis built his first sports car in 1923 or 1924. It was based on a Model T Ford and was supposed to resemble a Stutz, one of the sportiest production cars of its day. It could otherwise have been called a hot rod, but though no beauty, it was dressed up with a home-built speedster body and had a bit more class than the usual jalopy of its day. Kurtis called it his "Stuttering Stutz."

Through the end of the Roaring Twenties and the Depression years of the 1930s while other work occupied him, Kurtis said later, he had two hopes, building a winning Indianapolis racing car and producing his own custom sportster. (The term "sports car" did not become current until after World War II.) It was a dream that would take him 30 years to realize.

Many of the custom designs of Kurtis' early years were boat-tailed, like the popular and expensive Auburn Speedster. In 1940 he had a chance to refine his styling of a sports car. As related in chapter 2, Denver Ford dealer Billy Hughes had replaced the V-8 engine of a 1935 Miller-Ford Indianapolis car with a 255-cubic-inch Offenhauser. Hughes wanted to put the Ford engine into a sportster. Kurtis took a junked 1937 Ford frame and adapted 1939 Ford hydraulic brakes to it. Using a pair of junkyard Chevrolet fenders on a body that he designed himself, he produced a surprisingly modern-looking vehicle. Frank subdued his earlier boat-tail shape and produced for Hughes something that looked a great deal like the postwar British XK-120 Jaguar. The car was clearly a transition in Kurtis' artistic and styling sense toward the smoother, more rounded shapes of the postwar world.

For years Kurtis had collected low-mileage wrecks with sheet metal damage and turned them into his special creations, well before George Barris started work on customs in Los Angeles after World War II. In 1941 he picked up a wrecked Buick Century with only 500 miles on the odometer. The war intervened before he could work his magic on it, but in 1946 he hauled the car out of his storage shed and built a sleek, bright red custom body for it, slab-sided, with a look that would soon be seen on the 1949 Mercury and Lincoln production cars. The customized Buick had a steel-framed aluminum hard top, covered in canvas to mimic a convertible, a style which also would soon be appearing on cars from Detroit. Kurtis' top could be folded into three pieces and stowed in the trunk, an

Bill Stroppe, seen here at the wheel of his Kurtis 500S sports car at a West Coast road race, dominated the California SCCA and Cal Club seasons in 1953 and 1954. *Mark Dees collection*

arrangement that would later show up on the 1957–1959 Ford Skyliner convertibles.

In good Southern California style, the hood and trunk lids were "decked," with the releases operated from inside the car. The leather-upholstered foam rubber seats folded down to make a bed, like the Nash of that year, and Kurtis even made special aluminum luggage to fit the rear compartment.

As we have seen, Kurtis drove the Buick to Indianapolis in 1948 and it was such a hit that the attention bothered him. On the way home to Glendale, Kurtis rolled the car into a ditch in Nebraska to avoid a collision with another car and the hard top saved his life. So stout was the car that, with minor work, it was able to be driven back to California.

Frank Kurtis with his customized 1941 Buick, the car that created such a stir at Indianapolis and led to the Kurtis Sports Car and the Muntz Jet. *Arlen Kurtis Collection*

Frank Kurtis with the first Kurtis Sports Car in 1949. Though sporty in appearance, the KSC ran in very few races and was not as successful in competition as its successor, the KK500S model. *Arlen Kurtis Collection*

Automotive writer Tom McCahill saw the Kurtis Buick custom at Indianapolis and was entranced by the car. "Convertible (fit for) a King," he wrote in an October 1948 *Mechanix Illustrated* article. "It has speed, safety, luxury, and dazzling class." Kurtis repaired the Buick and sold it to Earl Muntz, a flamboyant Los Angeles used car dealer. Earl "Madman" Muntz ("I want to give 'em away, but Mrs. Muntz won't let me; she's CRAZY") was king of the hard-sell for cars, TV sets, car radios, motorcycles, and a dozen other postwar enterprises.

The reaction of McCahill and indeed throngs of spectators at Indianapolis intrigued Kurtis. With midget production declining as 1948 drew to a close, Kurtis considered building a sports car version of the Bill Hughes vehicle with a light engine, to be called the Kurtis Kar, but that idea died stillborn. The custom Buick offered a better prototype for a road vehicle, and Kurtis decided to build his own car along its lines, producing an attractive sports car that could be sold based on its style and the publicity his race cars had brought him. As an added advantage, the sports car would keep his shop busy when things were slow in the race car building business.

Kurtis planned to build a basic frame and body and let the customer install any engine he wanted and do the expensive, labor-intensive detail finishing work. His first Kurtis sports car emerged from the Glendale shop early in 1949. It was built as a two-seat convertible coupe. Kurtis planned to power it with a Studebaker V-8 engine that Harry Hartz, a Studebaker distributor, told him would soon be available from the factory. The engine was not ready when the car was finished, so the prototype was put together with a six-cylinder flathead Studebaker Champion engine while Kurtis waited for the V-8.

Not satisfied with the power the six produced, Kurtis added a McCullough supercharger to perk up the engine, and drove the car to Indianapolis in May 1949, hoping to generate the same excitement the Buick had created a year earlier. He was not disappointed by the car's reception at Indianapolis, but he was unhappy with the flathead Studebaker engine and Studebaker's delay in producing its promised V-8. Frank continued on to Dearborn after Indianapolis and was ushered in to see Edsel Ford, an amateur automotive stylist of no mean ability himself. Edsel was impressed with the KSC sports car and offered to sell him Ford engines and running gear. In fact, Edsel directed the managers at the Long Beach, California, Ford plant to sell Kurtis any parts he wanted to buy.

The new Kurtis sports car of 1949, as seen here, was based on Ford chassis parts with a body patterned after the customized Buick that Frank Kurtis built after World War II. *Arlen Kurtis collection*

Next, Kurtis arranged to take the sports car to Rosamond Dry Lake for a speed test. He was able to push it to 130 miles an hour, just short of the speed a Jaguar had reached earlier that year. In August, Kurtis took the car to the Bonneville Salt Flats in Utah, where *Hot Rod* magazine was holding the first Bonneville Speed Week. Wally Parks, *Hot Rod* contributor and head of the National Hot Rod Association, drove the

The frame of the 1949 Kurtis Sports Car as it was constructed at the Kurtis shop. As the photo shows, it was of conventional channel frame construction with a sheet metal pan. *Arlen Kurtis collection*

parked alongside a V-tailed Beech Bonanza, but sales were slow. Hollywood stars like Gary Cooper and Clark Gable bought Jaguars, not Kurtis Sports Cars. The Kurtis was not as sophisticated as its British competition and it was more expensive.

Kurtis soon realized that the car was not going to be the success he had hoped for. Mitsubishi showed interest in the design, but he refused to sell to them. Meanwhile Muntz was entranced by his Kurtis custom Buick. In late 1949, the Madman offered Frank $75,000 for the rights to the KSC car and Kurtis' tooling. Muntz actually bought the rights to the car in early 1950. There is some disparity over just how much he paid, the $75,000 Kurtis recalled or $200,000, a figure conjured up later by Muntz. After the sale, Kurtis built 10 bare chassis for Muntz to complete at his plant in Evanston, Illinois. Muntz added a rear seat, "much to our dislike," Kurtis said afterward. "He increased the size of the vehicle out of the sports class, used the then-current Lincoln engine and had such poor workmanship that it certainly did no justice to the car."

THE KK500S

Lieutenant General Curtis LeMay, head of the Strategic Air Command, was interested in racing. He opened several SAC air bases to sports car races, and at SAC headquarters at Offut Air Force Base near Omaha, he encouraged his crews to set up hobby shops where they could work on their cars. In 1952 LeMay flew to Los Angeles and spent a day with Kurtis, urging him to design an all-American sports car. The general took Kurtis along with him to visit some of the Southern California aircraft plants, and Kurtis got a big kick out of seeing the red carpet reception LeMay got from the contractors.

That summer, after the Indianapolis race, a friend offered Kurtis a ride in a British Allard sports car. The brutish Allards, powered by Ford, Cadillac, and Oldsmobile V-8s, were doing well in American sports car racing. Kurtis took the British car to Griffith Park to wring it out a bit. The Allard had a crude form of independent front suspension, using a Ford beam axle cut in two and pivoted at the center. Kurtis, while no racing driver, was certainly skilled behind the wheel. The Allard nearly got away from him in a high-speed turn and Kurtis, always the dreamer, told himself that he could build a better handling sports car.

In the early 1950s, only Briggs Cunningham was producing American sports racing cars, and the Cunningham production C-4 model was largely built to try to show the Internal Revenue Service he was running a real business, and to comply with the SCCA homologation rules. (Homologation rules in certain sports car classes are requirements that a certain number of a model be produced for public sale before they can be raced as a "production" car.) Kurtis sensed an opportunity and a new challenge. With Johnnie Parson's Indianapolis victory in 1950, Kurtis had achieved the pinnacle he had been striving for since he began building race cars, 20 years earlier. He had saturated the midget market. Now in 1952, his cars were the toast of the world, or at least the American racing world, but he had always loved building sporty road cars. The Kurtis-Kraft Indianapolis car had nearly won the race in 1952, bringing Kurtis-Kraft worldwide publicity. What better way to build a new sports car than to base it on the Indy car?

The 500S sports car and its kit cousin, the 500KK, announced January 16, 1953, were the result. The car Vukovich drove at Indianapolis had a basic tube frame and torsion bar suspension, and it handled very well at Indianapolis. The 500s were modeled directly on the Indy car and their handling was as positive and solid as anything that the Speedway had seen. Those cars were

This first version of the Kurtis 500S sports car, built for a Los Angeles Hudson dealer, had a flathead six-cylinder Hudson Hornet engine installed. The Hudson was not particularly powerful, and most subsequent versions had Ford V-8, Cadillac, Olds, or other far lighter, more powerful engines.
Johnny Pawl collection

the most successful of Kurtis' models in sports car racing and are the most recognizable today.

The 500S, following four years after the KSC and built as a competition vehicle, found itself a much better niche than the KSC. It was a two-seat sports racer, with no pretensions to luxury—an "Indy car with fenders." Rather than compete in the marketplace against luxury highway vehicles, the 500S was a race car. The design was strength and simplicity personified. Its handling and horsepower with the proper engine were outstanding, in what was becoming the American style of a big engine in a stout, lightweight chassis. Lance Reventlow and Carroll Shelby would soon carry the same type of chassis to the pinnacle of sports racing success.

Kurtis built the 500S to meet Federation Internationale de L'Automobile (FIA), the international racing organization, rules for sports cars and it became available in mid-1953. Drawings by John Bond, dated that October, show the basic Indianapolis frame, a combination of typical Kurtis hat-section flanged steel side beams and tubular cross-members. The distinctive grille of nine curved vertical chromed tubes was nearly identical to that of the Indianapolis car and, in fact, those tubes were stressed members of the frame. The 500S had solid axles and torsion bar suspension with an appearance much like that of the Allard, but with the handling ability of Vuky's roadster. Most versions of the 500S had cycle style fenders, though a few were given a clamshell-shaped front fender designed and built by George Barris. A planned closed car was never built, though a few buyers of the 500KK kit version installed coupe bodies on those frames.

Originally, brakes on the 500S were an early disc type made by the Catton Company of La Canada, California. The Catton was something of a transition between a drum type and modern brakes, using a lined disc squeezed between two plates. Joe Lencki had used a similar device on his 1946 Indianapolis car. The design was far less subject to fading than drum brakes. While the Catton brake was lighter than conventional cast-iron drum brakes, it was nowhere near as light nor as effective as the Halibrand disc and caliper brakes that were coming into use on racing cars. Nevertheless, Griff Borgeson wrote in *Hop Up Magazine* that the brakes on the Kurtis were remarkable. After a fast stop from 110 miles per hour Griff gingerly touched a drum and found it barely warm. In test stops from 60 miles per hour, the car averaged 127.5 feet, very good for the time. The Catton system, however, was soon outmoded and became a historical footnote. Most later 500 models used 1948 Ford drum brakes.

By February 1953, Kurtis had built a finished aluminum-bodied prototype with a 160-horsepower Twin H-Power Hudson Hornet engine for Joe Ricketts of Storey

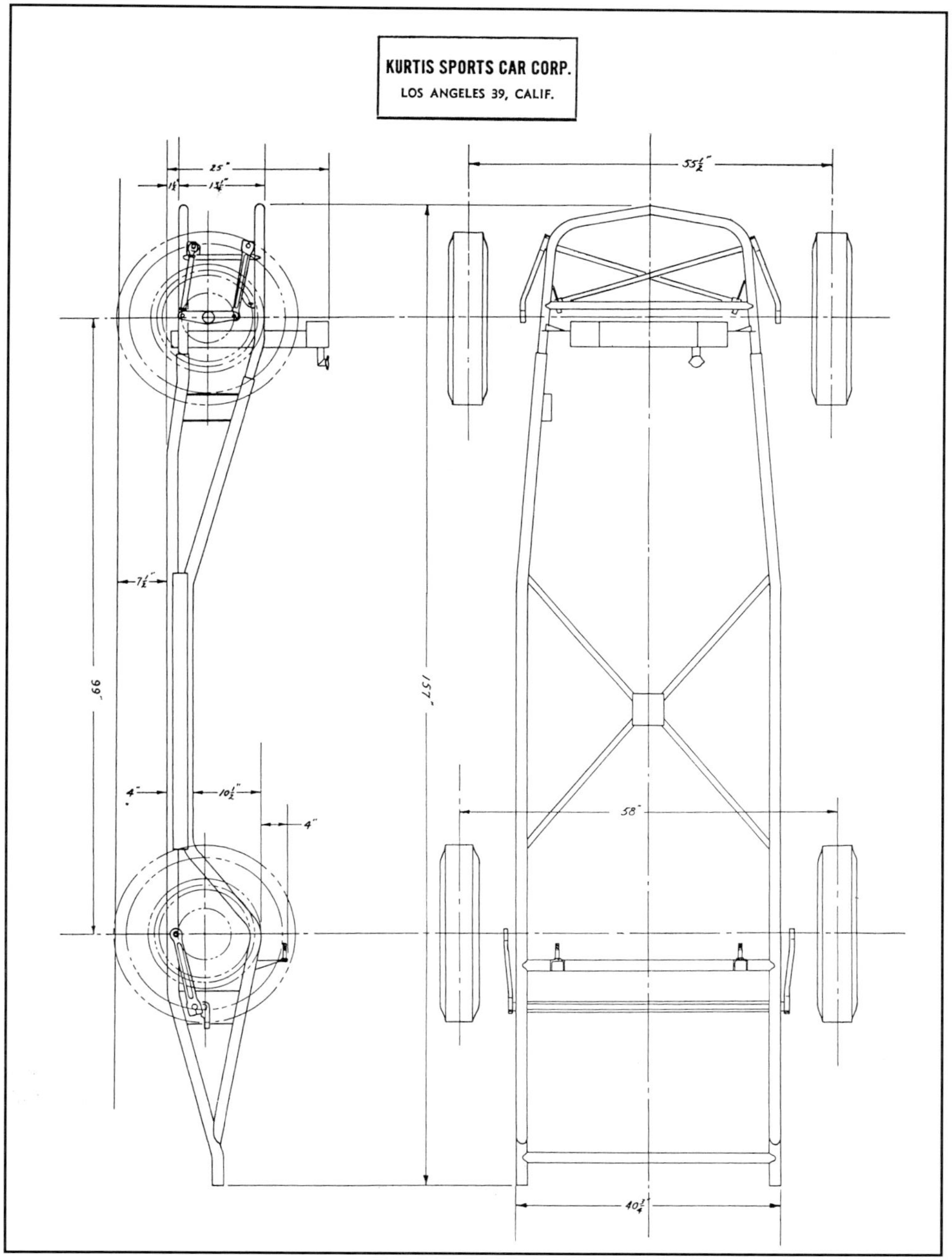

This is the frame drawing of the Kurtis 500KK kit car.
Leo Goossen Archive

Frank Kurtis examines the Catton disc brakes used on the first 500S sports cars. These were an early form of disc brakes that squeezed a lined disc between inside and outside steel plates. It was quickly outmoded by the Halibrand and other "spot" type brakes. The Catton design was built by a company in nearby La Canada, California. *Arlen Kurtis collection*

& Ricketts, a Los Angeles Hudson dealer. Some later versions had fiberglass noses and tails. All models had aluminum engine and cowl panels and steel doors.

Ricketts soon sold his 500S to Charles Painter, of Laguna Beach. *Road & Track* did a road test of the car that summer. The magazine found the Hudson engine too heavy and the gear ratios in the stock Hudson transmission unsuitable, but came away very favorably impressed with the car, despite the fact that its top speed was only about 103 miles an hour. "This is no Ferrari," the magazine said, "but it costs less than half as much and you don't have to bow low toward Milan to get parts." The Hudson-powered 500S weighed 2,265 pounds. Later versions trimmed off nearly 500 of those pounds.

Borgeson tested Frank McGurk's Cadillac-powered 325-horsepower 500S later in 1953 and found it much faster than the Hudson, capable of more than 150 miles per hour. Using a Hydra-matic transmission from a GMC pickup truck, the Kurtis went from 0 to 60 miles per hour in 4.7 seconds and did the standing quarter-mile in 13.6 seconds, faster by far than the Jaguar, the Ferrari America 4.1, or the then-hottest U.S. production car, the Olds 88.

Kurtis also offered a version, the 500KK, as a frame and running gear kit to which the buyer could add his own body and engine. Several firms in the Los Angeles area provided fiberglass bodies, including Glasspar, Woodhill, Allied, and Victress. Both the 500S and the kit car could be bought with wheelbases from 88 to 100 inches. (John Bond did the 500KK engineering drawing, which are dated Nov. 23, 1953.) At Indianapolis in May, the Belond KK500B had had welds break between the flat side beam and the tubular part of the frame, leading Kurtis to shift to an all-tube truss arrangement that was worked out that fall. After building 500S models for Stroppe and Frank McGurk, Kurtis redesigned the sports car to eliminate the hat-section side rails and replace them with a full tubular frame. The result was a slightly lighter, stiffer car.

As far as can be determined, 60 rolling chassis 500KKs were built, either bodied by Kurtis in aluminum and fiberglass or sold to customers who bought their own fiberglass bodies. The price for a car, less engine, was $4,586.39, still several dollars higher than the then current price for a Jaguar.

The 500 sports car made the road-holding ability of the Allard obsolete, and out-muscled the more refined Jaguar. At less than half the price of its nearest real competitor, the Ferrari, it was a winner on the track, though not a significant financial success for Kurtis. With the cloudy cash-flow accounting of the Kurtis company, it is hard to say more than it probably paid its way but made little profit.

BILL STROPPE

Kurtis donated a short-wheelbase 88-inch 500S chassis to Bill Stroppe of Long Beach, California, who installed a Clay Smith–tuned 200-horsepower Mercury V-8 and raced it in sports car competition in 1953 at Carrell Speedway, Palm Springs, Seattle, Del Mar, Moffett Field, Stockton, Chino, and Phoenix, winning four races of the five in which it started. The Stroppe car had special water pumps and plumbing to deal with the typical Ford overheating problem, one never really solved in the 19 years the flathead was produced. With Kurtis' adjustable torsion bar suspension, the car could be tuned to the track, and gear ratios could be changed easily in the Halibrand rear axle center section. The Kurtis could out-corner the Cunninghams and out-accelerate the Ferraris, which had to shift far more often than the stout and torquey Ford.

At Carrell in its first outing Stroppe and his Kurtis set fast time, and won the trophy dash and the main event. He had trouble handling the car on Carrell's dirt surface in the dash, but by the feature had gathered the car up a bit. If the Mercury V-8 had not threatened to overheat, he would have made the race a runaway. At the Palm Springs road race in March, Stroppe in the Kurtis had a wheel-to-wheel duel with Jack McAfee in a 4.1-liter Ferrari until his magneto quit.

The 32-year-old Stroppe and his mechanic, Chuck Daigh, took the Kurtis to Phoenix on May 3, 1953, and beat McAfee and another far more powerful Ferrari by three-quarters of a lap to win an 80-mile road course race. On July 12, 1953, Stroppe defeated a strong field of sports cars at Chino Airport, his third win in a row.

Later in 1953 Troy Ruttman, the 1952 Indianapolis winner, drove Stroppe's Kurtis at Riverside while Stroppe himself was running in the Mexican road race. The 6-foot 3-inch Ruttman did not fit well in the short car. The crew had to cut down the driver's side so he could hang his left knee outside the cockpit. He was running third against stiff competition from two Cunningham C4Rs and Jim Kimberly in a Ferrari when the Ford engine broke a rod. At March Field in 1953 Ruttman, more of a race track driver than a road racer, straightened the course a bit by knocking over the cones in controlled broadslides before the Ford broke a crankshaft. For the year Stroppe's Kurtis won seven of nine races it started, for a 0.777 percentage.

Frank Kurtis and Don Francisco of *Hot Rod* magazine look over Bill Stroppe's Mercury-powered 500S Kurtis sports car at the Chino Airport road races, July 12, 1953. Stroppe was an easy winner in the over-1,500-cc event. *Author's collection*

In 1954 Stroppe won the Cal Club's opening race at Santa Barbara and followed that up with firsts at Willow Springs, Moffett Field, and Sunnyvale. Stroppe and his Kurtis went on to win six straight races at Moffet Field, near San Jose, and the car was featured on the cover of *Motor World* magazine. He was by far the most successful driver of a 500S, though other well-known drivers did very well in them. With Johnny Mantz driving, the Stroppe 500S took second in an invitational race at Guadalajara, Mexico, despite a hay bale crash. When Mantz took out the hay bale Stroppe rushed up shouting,

Sam Parriott in his Kurtis 500S sports racing car. Parriott, a city councilman in City of Industry, California, and a already a grandfather, drove this car in NHRA sports drag races and in it he won a national drag racing championship at Indianapolis in 1955. *Jeff Parriott collection*

A/SP
146

"You've wrecked my car!" but the pair got the car back on the course in time to finish the race.

Though the hot rodders loved him, the sporty car crowd did not take well to Stroppe's victories in what they considered a lowly hot rod, an unsophisticated Indianapolis chassis with a 22-year-old flathead engine, but Bill kept on winning and ignoring the people whom the rodders described as "teabaggers." Soon enough the foreign iron would surpass the Kurtis in technology and bring the reign of the "specials" to an end, but not in the mid-1950s.

At Pebble Beach in 1954 Stroppe's Kurtis broke the track record by five seconds, and he won the main event. At Golden Gate Park in San Francisco, he finished third against the best West Coast Ferraris and was second at Torrey Pines to Ken Miles in the Troutman-Barnes Ford Special, while Chuck Daigh was third in a Kurtis-Lincoln.

Stroppe and Kurtis were readying an attack on eastern sports car racing in 1955 and even spoke of going to the 24-hour race at Le Mans, but Stroppe's partner, Clay Smith, was killed in a pit accident at the AAA Championship race at Du Quoin, Illinois, in August 1954. Stroppe had to sell the Kurtis to pay for Smith's funeral. The Internal Revenue Service padlocked the Stroppe shop for taxes, putting him out of business. Richard Seifried eventually bought the Stroppe Kurtis and won six vintage road races with it over five years. After passing through the hands of several more owners, John Queen and Lee Coker bought the remains, restored the car, and put it back into Stroppe's hands to vintage race in the late 1980s.

Murrell Belanger's Chrysler-powered Kurtis 500S, driven in the 1953 Carrera Panamericana road race in Mexico. Duane Carter took over the wheel after Tony Bettenhausen fell ill on the first leg of the race. It is seen here at Tuxla Guttierez with Carter at the wheel. *Watkins Glen Museum collection*

SAM PARRIOTT 500S

Sam Parriott, a city councilman in City of Industry, California, bought his own Kurtis 500S and had Bill Edwards stuff a supercharged Cadillac V-8 under the hood. Parriott was no stranger to speed. He had started racing on the dry lakes in 1924 and turned 137 miles per hour on a Crocker Twin cycle at Muroc in 1939. In 1954 he was ready to challenge records again, in sports drag racing and at Bonneville. Edwards added an S.C.O.T. Roots type blower to the Cad on a specially fabricated manifold. The chassis had a Halibrand quick-change rear axle, Chrysler brakes in Buick drums, and Halibrand knockoff hubs. Instruments included a manifold pressure gauge and a G-meter. Thus loaded for bear, Parriott quickly ran a 121–miles per hour pass in the quarter-mile at the Pomona drag races. Parriott would later have Kurtis build a more extreme drag racer, one of the last racing cars Kurtis produced.

MURPHY 500S

Bill Murphy, dealer for both Buick and DeSoto, bought two 500S models. The first, in 1953, was equipped with a DeSoto V-8, which, despite Chet Herbert's ministrations, was unreliable. Murphy was phasing out of the doomed DeSoto dealership, and his second 500S was equipped with a Buick engine. Murphy raced it from 1954 through 1955, with one of its best showings at Sebring in 1955, before a loose battery mount ended their day. The Murphy car was successful at Torrey Pines, Palm Springs, and other West Coast road races. It was distinguished from a standard Kurtis 500S by the clamshell fenders built for it by George Barris. The car was retired in 1955 and passed through Barry Williams' hands to Bill Chapin, who plans a complete restoration.

CARRERA PANAMERICANA

The Carrera Panamericana, the Mexican road race, was run from 1950 to 1954, a 1,900-mile high-speed grind from Textla Gutierrez on the Guatemalan border through Mexico City to Juarez at the U.S. end of the Mexican stretch of the Pan American Highway. The first

Bob Christie's Kurtis 500M. This car, equipped with a supercharged Nash Jetfire Six rocker arm engine, ran in the 1954 Carrera Panamericana road race. It kept up with the European sports cars in the first leg but the belt tensioner on the McCullough supercharger jammed in such a way that it overboosted the engine and melted several pistons. Christie, who also drove William Tucker's KK500J Kurtis at Indianapolis, said he was going more than 170 miles an hour when the engine let go. The car had a Nash driveline and rear suspension and a Kurtis frame, body, and front end.
Bob Christie collection

race was an open affair with few rules, but by 1952 it had become an international event. American Chuck Stevenson won the 1952 and 1953 races in a well-prepared Lincoln.

Four Americans, three of them (Murrell Belanger, Jack Ensley, and Ray Crawford) owners of Kurtis Indianapolis cars, put Kurtis 500S cars into the Carrera Panamericana in 1953. Tony Bettenhausen drove Belanger's Chrysler Hemi-powered car and Duane Carter piloted Leonard Fulkerson's Kurtis, which had a Chrysler Hemi V-8 engine. Bettenhausen was running with the leaders until a battery cable chafed through and started a fire.

Eventually the race wore out its welcome in Mexico and the 1955 tragedy at Le Mans put an end to any hope for its resumption.

AYULO 500S

Indianapolis driver Manuel Ayulo bought a Chrysler-powered KK500S in 1953 to run in the Carrera Panamericana. However, he was killed in a crash at Indianapolis, and the car was sold to John Fox, of Kentfield, California, who raced it in Cal Club (the California Sports Car Club, a competitor to SCCA, was the strongest sports racing club on the West Coast, beginning in 1947) and SCCA races in the West with a Cadillac V-8. Lou Brero, of Arcata, California, bought the car in 1955 and drove it for the first time at Santa Rosa Airport. That day, one end of the rear axle broke but, as it had a locked rear end, Brero could still race with only one rear wheel driving. Though Brero was able to stay close to race leader Sterling Edwards in a 3.5 Ferrari

Monza, the Kurtis was so loose in the corners he eventually dropped out.

Brero entered the car in local sports car drag events and though turning as high as 122 miles per hour in the quarter-mile, continued to twist off Ford axle shafts. In five races, the car went through six LaSalle transmissions, two Ford rear axles, and four engines.

Brero's son, Lou Jr., took the Kurtis to Ted Halibrand's Los Angeles shop and had a solid Championship car axle installed along with Halibrand disc brakes. The Cookes Machine shop hooked the Cadillac to a Jaguar four-speed transmission and did considerable work on the engine.

At the next race, at Seattle Seafair in Bremerton, Washington, in the rain, the brakes caught fire. After cutting enough holes in the cycle fenders to ventilate the brakes, Brero went out again to practice. At the end of the 4,700-foot back straightaway, he was making in excess of 140 miles per hour when another car, off-course, barreled back on to the track. Brero, in dodging the other car went into the woods, plowed through a forest of small saplings, and hit a stump that launched him across a 100-foot-wide taxiway and off into the woods again, upside down. When course marshals arrived, Brero was out of the car looking for the pieces it had shed on its way through the trees.

Repaired, the car raced a month later at Salinas and Buchanan Field, where Brero won his class. He took the Kurtis to Nassau for the Bahama Grand Prix in 1956. Brero Junior wiped out the bottom end bearings in the Cadillac in practice, but his father found a junkyard engine in Miami and had its crank, bearings, and rods flown to Nassau. In the Le Mans–type start, the Kurtis-Cadillac was the biggest engine there, and starting spots were chosen by engine displacement, putting it on the pole. Young Brero jumped off to a fairly spectacular early lead, dirt-tracking through the loose sand and coral for a lap and a half before the faster cars could find a way around him.

KURTIS SPORTS CARS SPECTACULAR EVERYWHERE

John Hammond bought a 500S, with a Dodge Hemi V-8 engine in 1953. He drove it at Pebble Beach in 1954 but went out with a sick engine. Dale Young then acquired the car and took it to Denver. At the Cheyenne, Wyoming, road race in 1955 with Walt McCormick at the wheel, the Kurtis finished second, and in 1956 at Cheyenne, McCormick won his class. Marvin Ruby drove it at a Pueblo, Colorado, road race, finishing third.

Young installed a 327 Chevy in the car in 1964, and Charlie Lowderman raced the car at Pikes Peak that year, finishing fifth and in 1966, fourth. A big-block, 366-cubic-inch Chevy went into the car in 1969, and Lowderman took another fourth with it at Pikes Peak.

Denny Weinberg and Tony Sampo, of Norwalk, California, owned another successful 500S, with a Chrysler Hemi engine in the car. Weinberg and Sampo drove it at Bonneville, going 173.41 miles per hour in the D-modified class in 1955. Troy Ruttman drove it on the dry lakes and in SCCA and Cal Club road races. In 1958, just after Ruttman returned from Monza, he pushed the car to the fastest sports car time of that year's Bonneville meet, nearly 180 miles an hour.

The ex-Frank McGurk Kurtis 500S sports car, later owned by collector Duncan Emmons. *Duncan Emmons collection*

Page 146
last paragraph
Denny Weinberg
is man I sold
my Kurtis to.
In 1959 Troy
Ruttman drove
the car at Bonneville
and clocked maybe
180 MPH for the
fastest sport car
time of that year

... drove his ... Class ... nship ... Cadil- ... urs of ... going ... meet ... e gro- ... urope ... wered ... here in ... quarters ... lman & ... vo 500S ... t ran in ... won the ... over Bill

... rtis 500S ... or) a cou- ... the Kurtis ... d the Fer- ... close com- ... Walters in ... d John Fitch

... cars with Of- ... He discussed ... d Dale Drake, ... ser engine to ... o and George ... jet engine was ... and had provi- ... and a battery ... experimenters ... o hire the work ... done, ... ity. As far as is known only Jack ... k delivery of an Offy-powered Kurtis.

Kurtis said in 1966 that about 21 of the 500S models were built, most of them as rolling chassis to which the buyers fitted engines of their choice. Customers included actor Keenan Wynn and Briggs Cunningham, who used his as a test vehicle for development of his own sports cars. Cunningham applied the 500KK front suspension design to his Cunningham C5R, helping cure a nagging tire wear problem. His earlier cars, C1 through C4R, had used independent front suspension, based (as were the Kurtis KK1000 and KK3000 Indy cars) on the 1949 Ford. At Le Mans, tire wear had been a very serious handicap for the C4Rs. In the 1953 Le Mans race the C5R used a Kurtis-style solid tubular axle with torsion bar suspension and leading arms. Years later, Cunningham told Kurtis at a reunion that John Fitch and Phil Walters ran the entire 1953 24-hour Le Mans race without a tire change, a dramatic improvement. (The C5R had a higher top speed than the 1953 winning Jaquars, but the Jag's disc brakes were far better than the Cunningham's drum models.)

One of the Kurtis 500S sports cars as restored in 1999 by Arlen Kurtis, Frank Kurtis' son, who has carried on his father's work by restoring Kurtis racing and sports cars and, indeed, building new Kurtis midgets.

PIKES PEAK RECORD

Kurtis 500S models raced in the Pikes Peak hill climb in 1959, 1964, and 1966, with a fourth in 1966 their best finish.

1959	KK500	Danny Jones	Kurtis	10th	8:10
1964	KK500	Charles Lowderman	Kurtis-Chevy	5th	6:21.5
1966	KK500	Bill Boldt	Kurtis-Pontiac	4th	6:26
	KK500	Charles Lowderman	Kurtis-Chevy	9th	6:33

A Kurtis 500M sports car as it was prepared to run in the 1988 re-enactment of the Carrera Panamericana. *Sexton collection*

This is noted collector Charles Davis' Kurtis 500M sports car as it ran in the Colorado Grande, a vintage racing rally held annually in Colorado. *Charles Davis collection*

The interior of the 500M Kurtis sports car demonstrates how Kurtis designed the car to be more of a street vehicle than a fire-breathing road racer like the 500S. *Mark L. Dees collection*

KK500M

The 500M sports car was the cause of the unravelling of Kurtis-Kraft Inc. in 1955. Conceived a year earlier to compete with the smaller, lighter, less expensive British imports and the Detroit-built Ford Thunderbird and Corvette, the 500M was a slightly smaller version of the 500, one tamed more to street use than the S model. The idea was attractive to Bill Allen, whose company had been doing tube bending for Kurtis-Kraft. Allen brought together a group of investors, including Robert McCullough, who offered to finance a company to build the car, to be designated 500 "M" for McCullough. The 500Ms were 9 inches shorter than the -S models and had all-fiberglass bodies produced by McCullough. The first prototype was equipped with a 172-cubic-inch four-cylinder Ford tractor engine of about 75 horsepower, making it eligible for racing in the smaller-displacement sports car classes.

The 500M was considered more attractive at the time than the 500S and the later 500X. As it was designed for street use, it offered a bit more comfort and room. To get more trunk space, the spare tire was mounted outside, Continental style. It was also built slightly more cheaply, with a stamped grille and rear fender scoops rather than the welded grille bars of the 500S. Aside from the scoop-shaped doors, the 500M looked a bit like a contemporary Ford Thunderbird two-seater.

Most of the M models were built to a wheelbase of 90 inches on a frame that was a shortened 500KK. It returned to the use of a Ford rear axle and other less-expensive parts, putting it in the same size class as the MG, Triumph, and Morgan. The weight, however, was up substantially, to 2,700 pounds. The tractor engine was no powerhouse, but of course a larger engine could be accommodated. With a Cadillac V-8, McCullough claimed the car would go 135 miles per hour.

A few did have fire-breathing V-8s installed, like one built for Ed Walsh, with an Olds Rocket 98 equipped with a McCullough supercharger. The M was actually quoted at $6,200 with a Cadillac V-8 and Hydra-matic. Two were modified to carry coupe bodies, which had much less wind resistance than the open roadster style body. One of those, Serial Number 35, was to have been used for the 1954 Mexican road race but was not completed in time. Robert McCullough had one built with a Cadillac V-8 engine.

One of the few 500Ms to compete in other than a stoplight drag was Bob Christie's Nash-powered Kurtis. Christie, of Medford, Oregon, ran it in the 1954 Carrera

Panamericana and was doing well until the supercharger overboosted the engine and melted several pistons.

McCullough was unable to turn out either good quality frames or bodies. The sports car business floundered and went bankrupt at the end of 1955.

Kurtis said in 1966 that he made 60 500 kit chassis, of which 18 were built as complete 500M chassis by the ill-fated Kurtis-McCullough Sports Car Company and were equipped with McCullough Motors Corporation fiberglass bodies. The rest had Glasspar, Victress, Allied, Woodhill, Sorrell, Byers, Bangert, or other fiberglass shells on them. Kurtis estimated in 1982 (probably optimistically) that another 50 or so were sold as parts.

KK500X

The Kurtis 500X was the competition successor to the 500M. Begun late in 1954, it was based on the then-current KK500C Indianapolis car. It had a slightly different front chassis layout from its predecessors, with the left suspension bracket on the axle allowed to rotate, while the right side was welded solidly in place. The frame was a fully tubular space truss, widened at the cockpit to meet FIA sports car regulations and with the battery and fuel tank as far aft as possible in the interest of weight distribution. The wheelbase was 96 inches. Steering, 3.2 turns lock to lock, was by Casale, and the 500Xs were equipped with a Halibrand quick-change rear axle and disc brakes. Ride height and handling were adjustable. Jaguar four-speed transmissions were used with Zidar adapters, though they often proved unable to take the power of the larger American V-8s. Weight, fully equipped, was about 2,200 pounds.

The body of the 500X was formed from aluminum panels that could be removed quickly with Dzus fasteners. The cycle fenders of the 500S were eliminated. The 500X was more rounded than the 500S had been and overall had a more modern appearance, though it had a door only on the passenger side. It was both lighter than the 500S and slightly more aerodynamic. Kurtis asked $6,000 for the car, less engine and transmission.

An estimated 12 500X cars were built. Six were sold as rolling chassis, without engines, and four as kits. Al Hoskins of El Monte, California, bought the first 500X, with a Cadillac V-8 engine and a Hydra-matic transmission, the driveline offset to the right by 3 inches. Jack Hinkle, of Wichita, Kansas, owner of two Kurtis Indianapolis cars, put a 91-cubic-inch Offenhauser midget engine in his, the second car finished, to compete with the Oscas in the SCCA's Class F Modified, under-1,500-cc, class. According to Kurtis, other buyers included racers Ak Miller and Jack Ensley. Bill Murphy, a Culver City Buick dealer, bought one—designated 500SX—with a Buick eight for power.

Jack McGrath, Hinkle's driver at Indianapolis in 1954 and 1955, did the Hinkle engine installation and chassis setup at the Kurtis plant and tested the car at Willow Springs, California. Hinkle was fairly successful with his Offy-powered Kurtis, winning several class events, including the SCCA event at Smartt Field in St. Charles, Missouri, on October 16, 1955. On the same day Ed Walsh won the Modified Production over-1,500-cc race in his Kurtis-Olds.

Mickey Thompson had a 500X with a 389-cubic-inch Cadillac engine, with which he turned 121 miles per hour within the quarter-mile. He advertised the car for sale in *Hot Rod* magazine in 1964 as capable of going

A Kurtis 500KK "kit" version of the sports car, equipped with a fiberglass coupe body by Allied Fiberglass Co., one of several companies that built plastic bodies in the 1950s. Allied and its competitors built several hundred bodies for kit and home built sports cars. *Arlen Kurtis collection*

This 500KK kit car is equipped with one of the Bangert Company's fiberglass bodies. *Bruce Trenery*

One of the last sports models Kurtis built was the 500X. The car seen here belonged to vintage racer Dan Bleidman at the time the photo was taken. *Harold Pace*

Another 500KK kit car, owned by John Furlow, a West Coast collector. The fiberglass body is by the Byers Co. *Harold Pace*

KURTIS SPORTS CAR BUYERS

William Rogers . . . . .Chicago . . . . .Studebaker
Austen James . . . . . .New York . . . . . .Ford V-8
Ed Walsh . . . . .St. Louis . . . . . . .Ford V-8
Walsh's brother in law St. Louis . . . . . . .Ford V-8
Jim White . . . . .Toledo . . . . .Chevrolet
Eugene Cassaroll . . .Detroit . . . . .Chrysler
Andy Granatelli . . . .Chicago . . . . . . .Ford V-8
Bob McCullough . . .Los Angeles . . . . .Cadillac
Juan Ismael . . . . . . .Manila, Philippines Ford V-8
John Young . . . . . . .Bakersfield, CA . .Ford V-8
Bob Wian . . . . .Glendale . . . . . . .Ford V-8
William Rogers . . . . .Chicago . . . . . . .Ford V-8
Rod Cameron . . . . . .Los Angeles . . . .Ford V-8
Unknown . . . . .Miami . . . . .Ford V-8
Arlen Kurtis . . . . . . .Glendale . . . . . . .Ford V-8
Eugene Cassaroll . . .Detroit . . . . .Kit
Unknown . . . . .Los Angeles . . . . .Kit
Unknown . . . . .San Francisco . . .Ford V-8
Earl Muntz . . . . .Los Angeles . . . . .Cadillac
(larger chassis) Total 21
Muntz Jet . . . . .Evanston, IL . . . .
. . . . .10 unfinished chassis

500 X Buyers:
001 . . . . .Albert Hoskins
of El Monte Cadillac later Thompson?
002 . . . . .Murphy, August 1955
003 . . . . .Hinkle Offy
004 . . . . .Zidar
005 . . . . .Chevy
006 . . . . .Missing
007
008 . . . . .Miller, Ford V-8Denver (Zidar)
Bob Schroeder, Chevy "21B/M"
Mickey Thompson, Pontiac
Breskovich, Cadillac

500 M Buyers
Unknown woman, Los Angeles . . . .
. . . . .short wheelbase, Ford 4-cylinder tractor engine
Milt Kahl, Los Angeles . . . . .Cadillac
Dr. Pease, Los Angeles . . . . .Cadillac
Bob Christie, Phoenix . . .Mexican road race, Nash
KK Demonstrator . . . . .Cadillac

500 S Buyers
Storey & Ricketts, Long Beach . . . .Hudson Six, later owned by Duncan Emmons
Unknown . . . . DeSoto Hemi
Jack Ensley, Indianapolis . . . . Cadillac
Bill Murphy, Los Angeles . . . . Buick, later owned by Bill Chapin
Bill Stroppe, Long Beach . . . . Mercury, later owned by John Queen and then Joe Sexton
Frank McGurk, Los Angeles . . . . Cadillac, later owned by Syd Silverman
Briggs Cunningham, Florida . . . . Cadillac
Murrell Belanger, Indiana . . . . Chrysler
Ray Crawford, Los Angeles . . . . (demonstrator)
Denny Weinberg, Los Angeles . . . . Chrysler
John Hammond, Monterey . . . Dodge Hemi V-8; ran Pikes Peak
Joseph Zidar, Denver
Bill Murphy . . . . DeSoto V-8, now owned by Duncan Emmons
Sam Parriott, Los Angeles . . . . Cadillac, now owned by Duncan Emmons
Mickey Thompson . . . . Cadillac
Mr. Buduron, Tucson . . . . Mercury
Keenan Wynn, Fresno . . . . DeSoto
Unknown, Pittsburgh . . . . Kit
Unknown, Chicago . . . . Kit
Ignacio Losanzo . . . . Hydra-matic Cadillac
Manny Ayulo, Los Angeles
Unknown, San Francisco . . . . Chrysler
Unknown, Las Vegas . . . . Mercury
Frank Sexton . . . . Carrera Panamerica
William Rogers

500X Cars
Hoskins / Mickey Thompson . . . . . . blown Cad
Hinkle . . . . .Offy
Murphy . . . . .Buick
Zidar Special
Nix . . . . .Chevy
Joe Sexton

190 miles per hour at Bonneville. Norm Babcock, of Santa Cruz, California, was another 500X owner with a 289-cubic-inch Pontiac V-8 in his car.

TIME MARCHES ON

While the 500X was a major chassis advance over the 500S, it did not fare as well in the marketplace. Its price of $6,000 was still above the Jaguar and other available sports cars. The Halibrand axle and brakes and Casale steering were lighter than the Ford parts of the 500S but cost much more; at the same time, the competition had caught up in the horsepower department. Finally, the mid-1950s were not a good market for exotic cars. The Jaguars, Porsches, and Ferraris were not selling well either.

The Kurtis 500S, -KK, and -M model sports cars went down the drain with the bankruptcy of the Kurtis-McCullough operation, but the day of the kit sports car was relatively brief in competitive racing. It seems likely that even had Kurtis not become involved with McCullough the sports cars would likely only have been produced for a year or two longer. As with almost any design in the racing world, one rules for a day before the competition catches up. However briefly, Frank Kurtis had made his mark in the transitory world of sports car racing. In the 1960s Frank would build a handful of one-off special racing cars. They, like his "500" sports models, are much desired today, in the twenty-first century, and several compete regularly in vintage racing.

MURPHY KURTIS-BUICK RACING RECORD

Willow Springs, 8/9/55 . . . . .testing
Torrey Pines, 10/22/55 . . . .dnf in 6 hour enduro (radius arm broke)
Torrey Pines, 10/23/55 . . . . .1st
Glendale, 11/13/55 . . . .1st in class, 4th overall
Palm Springs, 12/3/55 . .1st in class, 2nd overall
Palm Springs, 12/4/55 . .1st in class, 4th overall
Torrey Pines, 1/14/56 . . . . .dnf (water hose)
Torrey Pines, 1/15/56 . . . . .dnf (spun while 3rd)
Palm Springs, 2/25/56 . .1st in class, 3rd overall
Santa Barbara, 3/17/56 .1st in class, 5th overall
Santa Barbara, 3/18/56 . .1st in class, 2nd overall
Pebble Beach, 4/22/56 . . . . . . .dnf (water hose)
Pomona, 6/23/56 . . . . .1st overall
Pomona, 6/24/56 . . . . .1st overall
Santa Maria, 7/8/56 . . . . .1st Overall
San Diego, 7/21/56 . . . . .1st overall
San Diego, 7/22/56 . . . . .1st overall
Santa Barbara, 9/1/56 . . . . .2nd overall
Sacramento, 9/30/56 . . . . .1st overall
Pomona, 10/20/56 . . . . .1st overall
Pomona, 10/21/56 . . . . .dnf (clutch)
Palm Springs, 11/17/56 . . . . .dnf (brakes)
Pomona, 1/19/57 . . . . . .1st in class, 2nd overall
Pomona, 1/20/57 . . . . .dnf (engine)
Santa Barbara, 4/6/57 . . . . .12th
Santa Barbara, 5/19/57 . . . . .dnf (piston)
Pomona, 7/28/57 . . . . .dnf (axle broke, C. Daigh, driver)
Riverside, 9/21/57 . . . . .2nd in class, 4th overall
Riverside, 9/22/57 . . . . .1st in class, 2nd overall

8/79 Murphy sold to Richard Seifried. Car was run subsequently in vintage eventsEND

THE MUNTZ JET

When Earl Muntz saw the Kurtis custom Buick in 1949 he was entranced. The brash, loud-talking used car salesman had never seen a car he loved more. When Kurtis developed a production version, the Kurtis Sports Car, Muntz bought two and knew immediately that he could sell them. When Kurtis realized he that as good as he was at building cars, he was hardly a salesman in Muntz' class, the Madman stepped up and bought the line, tools, rights, and unfinished chassis.

After all, Muntz had prospered pushing television sets, radios, stereo tape decks, even motor homes, and $72 million dollars worth of used cars in the single year of 1947. He could sell ANYTHING. Of course, anyone could sell cars in 1947, when the wartime shortages were still being made good. It was one time when the master salesman let his heart rule his head.

But in the fall of 1950 the Kurtis Sports Car looked like a golden opportunity. Imported sports vehicles were still rare and glamorous objects and America had not produced anything to match them. The Thunderbird and the Corvette were still in the future and there was a demand to be met. A certain amount of madness had surrounded Preston Tucker's Tucker Torpedo until it proved to be far more hype than reality. The Kurtis was being produced. Muntz was certain he could do it better.

Earl said he paid Kurtis $200,000 for the design, tooling and rights to produce the sports car, while Kurtis recalled receiving only $75,000. The name was changed to the "Muntz Road Jet." The Madman also felt the car would be more salable as a four-seat sports convertible sedan. His and Kurtis' accounts differ there, too, but according to Muntz, Kurtis helped him design a stretched version, 13 inches longer, to accommodate a back seat. The Muntz was also given a removable padded steel top, a bit cheaper than the hand-fitted folding aluminum top on the KSC. Kurtis said later that the longer wheelbase spoiled the car. Despite his opinion of the modifications, he and Sam Hanks worked with Muntz in preparing the new car and Frank later drove a Muntz (which was heavier and slower than the two-seat Kurtis) 128 miles per hour at Bonneville in an effort to boost its reputation as a speedster.

The Muntz Jet had a wheelbase of 113 inches and weighed 3,500 pounds. It was powered at first by a Cadillac V-8 engine, though after the first 28 cars were built Muntz went to a heavier Lincoln powerplant. The plant in Evanston, Illinois, was not designed for economical production. With too much hand work in the car it cost more to produce than it could be sold for.

Designed for the high-priced market, it listed at $5,200, a price higher than most contemporary Cadillacs and Lincolns. It was a tough sell. Muntz' own personna probably hurt sales to the more wealthy people the car was designed for. In 1954, after building 394 cars, Earl reluctantly shut it down.

Muntz said he lost $1,000 on every one of the cars. The current Mrs. Muntz must have been certain the Madman was crazy.

A CAR SPOTTER'S GUIDE TO KURTIS SPORTS CARS

Jack Ensley's Kurtis 500S was equipped with a big 430-ci Lincoln V-8 engine. It is seen here finishing a leg of the Carrera Panamericana at Oaxaco, Mexico in 1953. *Watkins Glen Racing Museum*

Frank Kurtis designed and built six sports car models, the KSC or Kurtis Sports Car, the Muntz Jet, the 500S, 500KK, 500M, and 500X, of which there was a 500SX variant. He also built two cars in 1963 and 1964 that, while they used similar frames, had significantly different bodies.

Kurtis Sports Car: This 1948 car was based on Frank Kurtis' customized Buick convertible. It had only a front seat, but it had a removable fabric-covered top. Suspension was essentially from the 1949 Ford, with coil spring independent suspension in front and Hotchkiss-style semielliptic springs in back.

Muntz Jet: Kurtis, with the help of Sam Hanks, redesigned his KSC in 1950 for Earl "Madman" Muntz, to whom he sold the tooling and rights to the sports car. Originally designed for the Cadillac V-8 engine, it had front and back seats, and most models were equipped with Lincoln V-12 engines. Suspension was identical to the KSC.

500S: This was the first competition Kurtis sports car. It was patterned on the Kurtis roadster that won the Indianapolis 500 in 1953. It had cycle fenders and could be equipped with a variety of V-8 engines. Suspension was via torsion bars and both axles were solid.

500KK: This was essentially the same frame and running gear as the 500S, sold as a kit for the buyer to mount any of many fiberglass bodies and the engine of his choice.

Pete Hasmer's 500KK kit sports car is equipped with another fiberglass body. This car was prepared for one of the re-enactments of the original Carrera Panamerica. *Harold Pace*

The Kurtis 500M sports car of 1954 was the third in the line of sports cars to be produced by Kurtis since 1949. This fiberglass-bodied roadster was to be turned out by the McCullough Company, but few were built before McCullough collapsed in 1955. *Author's collection*

500M: M for McCullough, an ill-fated Kurtis venture. This was a street sports car with a slightly more streamlined-appearing body. Similar to the 500S under the skin.

500X: This was a more elaborate tube truss frame car with a full body that covered the wheels. Solid axles and torsion bar springs. Had a Halibrand champ car rear axle and disc brakes. Intended for racing only.

328 AHSR
520
City of Industry
Special

Kurtis in Eclipse

CHAPTER 6

Kurtis cars did not win at Indianapolis after 1955, though they remained competitive, finishing in the top 10 through 1962. If multicar crashes in 1958, 1959, and 1961 had not taken 11 of his fastest cars out of the race, his record undoubtedly would have been better. At the end of 1955, Frank Kurtis' world was in a shambles. The midgets with which he had made his reputation had declined drastically in popularity, and he had allowed Kurtis-Kraft to become entwined in a business deal with Robert McCullough that almost ruined him. As 1956 dawned, his sports car ventures had collapsed, and he was even forced out of his shop on Alger Street.

It is difficult to tell just why one racing car succeeds and another does not. The rear-engine revolution at Indianapolis in the 1960s was a clear technological change, as had been the Kurtis roadster, or offset car in 1952. The roadsters outmoded the upright dirt cars early in the decade because they were lower and had better weight distribution for turning left on oval tracks. The Kurtis 500 was the first modern, well-designed, offset car. But the differences between the roadster models that raced from 1952 through 1964 were less clear.

For a dozen years, the winning cars had in common only the fact that their engines were offset from the centerline. They had laydown engines and upright engines and layover engines. All used the Kurtis-style torsion bar suspension with solid axles, but the bars were arranged in different ways. There were leading link, trailing link, and watts linkage arrangements and cars with any of these setups were lapping Indianapolis within fractions of a second of each other, certainly within the margin where driver skill probably made the difference.

There was conventional wisdom that independent suspension did not work on dirt, yet Johnnie Parsons won the AAA National Championship in 1949 in the independent suspension *Kurtis-Kraft Special*, against such ferocious competitors as Rex Mays, Bill Holland, Tony Bettenhausen (in the Belanger Number 99 no less), Joie Chitwood, Myron Fohr, and Emil Andres. In fact the *Kurtis-Kraft Special* was the second-most winning car Kurtis ever built. The solid axle Number 99 car was Kurtis' absolute best, and it won at Indianapolis and 15 times on the dirt.

The answer to the decline of fortune for Frank Kurtis–built cars at Indianapolis owes much to luck—crashes and engine failures—but probably, as A. J. Watson observed, "Race cars are like clothes. People buy in fads. Kurtis cars were the fad, then mine were." The Kurtis fad at Indianapolis was nearly over after 1955.

Sam Parriott's 1963 Kurtis sports drag car forges ahead of a Ford coupe in an AA/sports modified drag race at Pomona, California in 1964. *Jeff Parriott collection*

For several reasons, Kurtis was on a downward slope. Perhaps too many owners who bought Kurtis cars were wealthy enough to own Indianapolis cars but not expert enough to run them competently. The one consistent fact is that when Kurtis himself was involved with the Kurtis-Kraft Special, or the really best mechanics were on a team, as with the Belanger Number 99 and the Vukovich cars, they won.

The McCullough arrangement came about when Kurtis saw it as a way to get out of his partnership with Walsh. A shy man, Frank had never been able to bring up his dissatisfaction directly with Ed. Years later he said "There may be a day when I can bring this up with him without dissension, but if I am so stupid [as] to get myself involved, and am so trusting, I've got nobody to blame but myself."

In 1956, as the McCullough debacle deepened, Frank was able to build only five Indianapolis cars, the smallest production for the shop since 1947. His son, Arlen, returned from the Navy in the winter of 1956 to find chaos. Arlen helped move the remains of the business to Earl Farand's old Indian motorcycle shop on Colorado Boulevard (later Colorado Street).

Although he built 14 of the KK500G model Indianapolis cars in 1957, the national economy was in decline in 1958 (". . . a recession to curl your hair," according to Treasury Secretary George Humphrey). Race sponsorship money dried up. Frustrated with the car owners, he would build only 5 more Championship cars in his lifetime. Twenty-five years later he would say "I got a lot more appreciation from Lockheed than from race people." He would build small racing cars for children, go-karts, and aircraft support machinery but never again would he see the glory years of the early 1950s.

After his disastrous partnerships with Ed Walsh and the McCullough group, Kurtis, so skilled at designing and building racing cars, was forced to start over again, because he was less adept at business and at the skills needed to keep his name before the public. He was an old world man, at home where he ruled his household, but he was shy when facing the outside world. His workers described him as pleasant at the shop, even sometimes a practical joker. He was known as a boss who occasionally even cleaned the plant washroom.

He was happiest when working on a new project. He was able to visualize a race car and to sketch it on paper (John Bond, Avard Ward, and Carl Newell drafted the final blueprints) and to design an entire passenger or racing car, a task that in Detroit took teams of many engineers. Usually his cars were successful, many of them technologically brilliant. But more worldly bosses, clients, and customers—particularly wealthy customers—intimidated him. He shied from confrontations. He was equally shy with the public at large and with the press, willing as one observer said, to "let his tin snips do the talking for him." He seemed unable to share the camaraderie of the racing people and the reporters who covered them.

His inability to be "one of the boys" with other racing people and the press gave him a standoffish reputation with the magazine and newspaper writers. "I was too busy to go to the cocktail parties," Kurtis would say. Actually he was too shy to mix easily with men who had better educations and more money. At the same time, Kurtis craved credit—if not attention—for his achievements. He was unable to shrug off what he felt were slights from the racing press. When Frankie Del Roy said in print that he had "built" the Parsons Indy car and the successful Buck Wheeler Racing Team Kurtis midgets, Kurtis was outraged. Though it was common for mechanics to say they had "built" a car when they had only installed the engine, painted it, and done the final setup, Kurtis felt they were taking too much of his credit.

What Kurtis saw as extreme conservatism of the racing establishment, including most drivers and owners, bothered him. He was criticized in 1948 when a part broke on the *Kurtis-Kraft Special* the first time out, and for a while it was hard to get a driver to trust it. In 1950 he rebuilt Nat Rounds' rear-engine Indianapolis car, giving it a suspension system much like that of his later very successful roadsters. He related that a "top-name driver took the car out at Indy and after just one lap came in saying he couldn't get the feel of it. After that no one else would give the car a chance."

"Eventually many of my ideas were accepted," Kurtis said, "but the thing that discouraged me was that it never got any easier. I would have just as much trouble trying to introduce something new today (in 1960) as I had in 1939."

Compared to Kurtis, A. J. Watson, who took the Indianapolis crown from him, kept his ego in check. Watson and Kurtis were both superb mechanics, and when Kurtis was deeply involved with the first *Kurtis-Kraft Special*, in 1948, 1949, and 1950, the car won the AAA National Championship and the Indianapolis 500. Thereafter Kurtis stepped aside from active participation in the racing of his cars to concentrate on car building. Watson, a more involved chief wrench, took Kurtis' designs and improved them. In later years A. J. quite cheerfully acknowledged that Kurtis set the standard for racing cars in the 1950s and that he, Watson, just made them better.

In 1970, by then retired for half a dozen years and still fuming inwardly over the real and perceived slights he felt he had received from the racing community and the press, Kurtis could stand it no longer. He had met Edward Hitze Sr. at Indianapolis and Hitze, who did restoration work for the Speedway Museum, was an avid Kurtis fan. Somehow Kurtis and Hitze came up with the idea of doing a book settling Kurtis' scores. In February 1972, Kurtis wrote Hitze "maybe we better go ahead" with the book. Kurtis made it clear that one aim would be to "rebuke that . . . Granatelli," whose own 1969 book, *They Call Me Mr. 500*, had dredged up the old bad feelings between the two men. Kurtis told Hitze that he would pay him for writing the book and give him "[a] share of the profits, if any. I'm in fairly good shape financially," he added, "but I'm far from a millionaire."

Over the next two years Kurtis sent Hitze a file of photos and a long series of hand-written reminiscences of his early life and his career as a race car builder. The resulting book, *The Kurtis-Kraft Story*, was published in 1974.

Kurtis did enjoy the annual Gilmore Roars Again party, thrown each summer by Gordon Schroeder. He had a grand time seeing old friends there, at the Gilmore adobe home behind the CBS Television Studios, where the Gilmore track had been. Despite his personal shyness, he almost never failed to answer letters from his fans, particularly, late in his life, from those who had bought old Kurtis race cars and were restoring them to their former glory.

By 1960 Frank Kurtis Inc. was doing very well building Air Force starter carts and quarter-midgets. Lockheed, the starter cart customer, was very pleased with the Kurtis product, particularly at the prices Kurtis charged. The result was that Kurtis largely withdrew from Indianapolis, leaving a gap that Watson, Floyd Trevis, Ed Kuzma, Lujie Lesovsky, Wayne Ewing, Wally Meskowski, Quinn Epperly, and a few others filled.

Kurtis continued to enjoy designing and building innovative cars. He returned to Indianapolis in 1963 with the John Chalik car but was deeply disappointed with Chalik's insistence on using battery ignition, which cost its Offenhauser engine at least 75 horsepower. The car had more aerodynamic drag than other cars of the era because of the roll cage built into it for Junior Johnson, and its suspension was too soft. Like most racing cars, it needed development and there was far too little time to do that. It is amazing that Kurtis racing cars often went so well without extensive testing, and that fact alone was a testament to their basic simplicity.

The final days of Frank Kurtis' race car work came in the early 1960s when he turned his energies and his innovative mind to advanced sports cars, dragsters, and cars to race at Bonneville.

THE LAST KURTIS RACING CARS

Even as he turned more and more to the aircraft companies for his business, Kurtis still liked to draw designs for race cars. When old friends asked him to work on their cars he couldn't refuse. Buzz Lowe, who had helped Kurtis find the materials to build his first postwar midget in 1945, asked him to rebuild a Ford Model A pickup truck for straightaway running on the dry lakes and at Bonneville. In 1962, in addition to the work for Lowe, Kurtis put together a special sports car Ed Walsh ordered to carry a three-cylinder Saab engine in H-Modified SCCA events.

Nick Colbert with his Kurtis-built dragster at the Continental Divide Raceway near Castle Rock, Colorado. This was the first AA dragster Kurtis built, though he later put together sports drag racing cars for Sam Parriott, Jack Lufkin, and others. The only other slingshot-style dragster he built was for Ron Phelps, of Pasadena, California. Though the Kurtis sports dragsters were very successful, the Colbert and Phelps cars were never worldbeaters.
Nick Colbert collection

Opposite page
Dean Lowe's A roadster pickup dragster and dry lakes car. Though this looks like a 1929 Ford Model A truck, in reality it has a Kurtis Indianapolis type frame under it and a Kurtis torsion bar front suspension. Lowe bought it from Joe Gemsa in 1959 and learned to drive in it. In 1960 Frank Kurtis rebuilt the chassis. Dean's father, Buzz Lowe, and Jack Lufkin built the 283-cubic-inch Chevy V-8 engine, and it ran in the NHRA B street roadster class. It held NHRA class records from 1961 through 1966 and was a class winner for five years. The car had a Chevy four-speed transmission and a Halibrand quick-change rear axle. It was polished off with 30 coats of high-gloss Candy Persimmon lacquer by Junior Conway. *Dean Lowe collection*

In the spring of 1962 Andy Granatelli ordered two new rear-drive chassis for the Novi engines he had bought from Lew Welch, and Kurtis could not resist building the new Novis in a frenzy of work before the race at Indianapolis. Kurtis called them his 500K models. Granatelli, he said, stiffed him for the payment on the last Novi, further souring his view of Indianapolis, but he would always try his hand at designing an innovative car for a buyer who wanted something new.

John Chalik, the Indiana plastics manufacturer, was the last straw for Kurtis at the Speedway. The car Kurtis built for him was an innovative chassis that suffered from a driver who walked away from Indianapolis after a few laps, and a stubborn owner, unwilling to listen to advice from his mechanics. It was the last Indianapolis car Frank Kurtis would build.

The fourth Kurtis car of 1962 was a slingshot dragster built for Nicholas Colbert, son of Lester Colbert, then president of the Chrysler Corporation. "Tex" Colbert was a trustee of the Automotive Safety Foundation, and if his son, Nicholas, was going to drag race, he wanted him to be in a professionally built car, not a home-built.

In 1960 Nick Colbert was a college senior at the University of Denver, letting off his enthusiasm for speed by racing dragsters at the Continental Divide Raceway. He had been national AHRA champion in the SS/Stock class in a Dodge D-500-1 in 1956. Gene Cassaroll, head of Auto Shippers Inc., was a long-time Indianapolis car owner and a subcontractor to Chrysler. His trucks carried most of Chrysler's passenger car output. Gene and Lester were good friends, and Cassaroll suggested Kurtis build Nick a chassis for the blown top gas eliminator class.

Young Colbert talked to Kurtis, who had never before built a dragster, and thought Kurtis had interesting ideas that might help him set records on the quarter-miles. He paid a deposit, and Kurtis went to work. Kurtis went to two drag events to see what the cars looked like. The hottest ticket at the time was a slingshot-design car built by Jack Chrisman of Willowbrook, California. Kurtis and Colbert decided to take the Chrisman design and improve on it. Kurtis drew up a relatively short chassis with torsion bar front suspension.

The Kennedy administration was just then being challenged in the Cold War arena by Nikita Khrushchev, who threatened to cut off Allied access to Berlin, as the Russians had done 10 years earlier. Kennedy reacted by calling up the Army Reserve and the National Guard. Colbert, a reserve officer, was ordered to active duty. Colbert spent six months in uniform while the car was being built, unable to watch it go together or confer with the shop on the design. With Colbert unavailable, Kurtis was in no hurry to finish it. The car was finally delivered in the spring of 1962. It had Kurtis cross-torsion bar suspension in the front and a solid axle-to-frame setup at the rear. It was equipped with a 470-cubic-inch 1963 Plymouth wedge V-8 engine with a C-T Automotive stroker crank, carrying a 6-71 GM Roots blower.

The car's best elapsed time was 8.56 seconds for the quarter-mile at 178 miles per hour at Half Moon Bay, California.

In Colbert's recollection, years later, the expensive car did not work well. Strongly built, it was too heavy by drag racing standards. Colbert had begun operating the Continental Divide strip just outside Denver and was too busy to test the car thoroughly, or, indeed, to run it much. When the car did run, it was not fast. Colbert missed having the chassis builder help develop the car. Kurtis often went to Indianapolis to see what was going on with his cars but not to drag strips in Colorado. "I got no help from Kurtis," Colbert said years later. Kurtis, of course, had no drag racing experience and built only one other slingshot-style car.

In 1963 Colbert built up an even larger Dodge wedge V-8 engine of 520 cubic inches, but was still disappointed by his results with the Kurtis chassis. Finally in 1964 he went to Don Garlits and had Big Daddy build a lighter chassis for the big V-8. It did better. In fact Colbert remembers the new car setting top speed at its first meet. The new chassis was, he said, half the weight of the Kurtis. The Kurtis chassis was sold to a racer in Omaha, Nebraska, and passed out of sight.

In 1963 another budding drag racer came to Kurtis for a chassis. Ronnie Phelps, son of a wealthy Pasadena family, decided he wanted to try dragging. When Ronnie was 12, his parents had bought him the second Kurtis-built three-quarter-midget, a little car he raced with Buzz Lowe's son, Dean. In 1963 Phelps was ready to step up to the big time. Kurtis, no engine builder, sent Phelps to Joe Reath, in Signal Hill to build a Chevy V-8 engine for the car. Reath said, many years later, that he called Kurtis and asked him, "What are you doing building drag cars?" Kurtis replied, "These folks have bought midgets and TQs from us. They want only the best. Money is no object. If they want us to build a dragster we'll give them the best we can."

Having done the Colbert car in 1961–1962, Kurtis was able to come up with a lighter design for Phelps, this time a rear-engine car. It was completed in mid-1963, and Phelps raced it for about two years with indifferent results. No record of its racing history has survived.

BUZZ LOWE CARS

Kurtis' oldest friends asked him to work on their cars, and that kept his hand in the car-building business to some degree.

In addition to the pickup truck, Buzz Lowe had had Kurtis slightly modify a stock Studebaker Starlight coupe for the dry lakes, but his son, Dean, rolled the car at El Mirage and totaled it, breaking Dean's back and putting him in the hospital for several months. Thinking that the Studebaker independent front suspension was not good for high-speed straightaway runs, Lowe took a second Studebaker coupe to Frank for more serious frame work. Kurtis put a solid front axle in the car, suspended on crossed torsion bars. It was far more successful.

SERIAL NUMBERS OF THE LAST KURTIS RACERS

Gerhardt 500-J Indy Car	C-121-61
Vatis 500-J Indy Car	C-122-61
Tucker 500-J Indy Car	C-123-61
Lowe Pickup	D-1-62
Walsh sports car	H-S-sports
Dale Burt (Aguila) sports car	62-S-1
Paxton 500-K Novi	C-124-62
Paxton 500-K Novi	C-125-62
Colbert dragster	D-2-62
Lowe Studebaker	D-3-62
Lufkin Corvette	D-4-62
Phelps dragster	D-5-63
Chalik 500-L Indy car	C-126-63
City of Industry drag car	S-D-6-63
Lufkin streamliner	S-D-64 (Last Kurtis racing car)

The engine of a Kurtis quarter-midget. Many of these small cars used different engines of the lawnmower variety made by West Bend and Kohler. Later Yamaha engines of 150cc and 250cc were used, some even burning nitromethane fuel mixtures. *Arlen Kurtis collection*

Young Mike Randall with trophies he won with a Kurtis quarter-midget in 1957. *Arlen Kurtis collection*

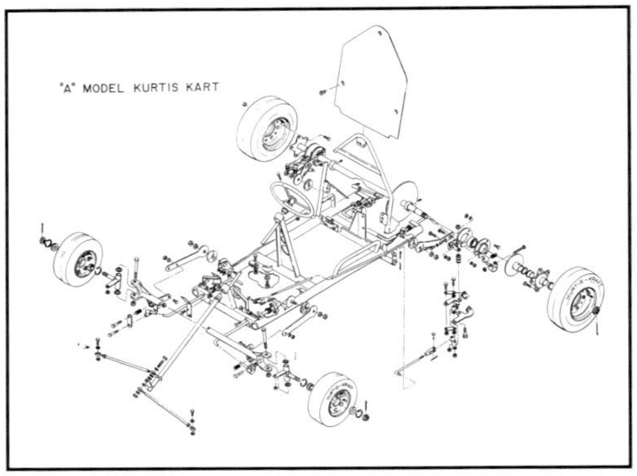

This was the diagram for Frank Kurtis' "A" model Go-Kart. *Arlen Kurtis collection*

A Kurtis quarter-midget of 1955, offset like the Indianapolis cars. *Arlen Kurtis collection*

ROCKET SLEDS AND STARTER CARTS

Frank Kurtis Inc. was making a tidy living from the aircraft starters it built for the North American Aviation F-100 Super Sabre jet fighter. In Kurtis' words, "Who needed Indianapolis with their cheap customers and endless rules?" North American was delighted to have a practical metal cutter and bender working for it rather than another aerospace subcontractor, top-heavy with analysts and overhead. The aerospace firm signed new contracts for starter carts in 1963, buying 26 in the following three years, along with 16 other pieces of equipment, business that kept Kurtis prosperous. Among special projects that Kurtis took on for the government contractors was a prototype of the Forsythe experimental all-terrain vehicle, for Lockheed, and the Terrastar, an Army amphibious vehicle.

Colonel John Stapp had ridden a rocket sled 632 miles an hour at Holloman Air Force Base at Alamogordo, New Mexico, testing ejection seats for high-speed aircraft, work that resulted in the design of the familiar Martin-Baker seat used in most military aircraft. The most spectacular vehicle Kurtis built for North American Aviation was a rocket sled used for parachute testing at Edwards Air Force Base. North American's program at Edwards reached even higher speeds than had been attained at Alamogordo. Cook Research Company, of Chicago, was the subcontractor, but when time came to build the sled itself, Kurtis was called in.

North American designed and built the propellant tanks, frame, and undercarriage, while Kurtis put the skin on the vehicle. North American's Rocketdyne Division provided the rocket engine that burned $350 worth of methanol and oxygen on each run.

The Cook Research Corporation Parachute Test Vehicle (PTV), a 25-foot-long rocket sled, built by Kurtis in 1952 for the Rocketdyne Division of North American Aircraft Corporation. This was the fastest vehicle Kurtis ever built. According to the U.S. Air Force, it attained a speed of 886 miles per hour. The sled was used on 123 runs between October 28, 1952, and May 1956 to test parachutes for missile warheads. *Air Force Flight Test Center*

The Kurtis rocket-propelled PTV throws up a cloud of spray as it enters the water brake at the end of the 10,000-foot Experimental High Speed Track at Edwards Air Force Base. The PTV was powered by a 50,000-pound-thrust Rocketdyne liquid-fueled rocket engine, using liquid oxygen and alcohol. It reached a speed of Mach 1.2 in 4.5 seconds. *Air Force Flight Test Center*

The vehicle was 25-1/2 feet long and 5 feet tall, propelled by a 50,000-pound-thrust liquid fuel rocket. Top-secret at the time, the program was testing reentry 'chutes for the warheads of Jupiter Intermediate Range Ballistic Missiles. The Kurtis-Cook rocket never carried a human cargo, but between 1952 and 1956 it made 123 unmanned runs at Edwards, achieving a top speed of 886 miles per hour, probably the first land vehicle to exceed the speed of sound.

In the late 1960s, Lockheed turned to Kurtis to build starters for its supersecret SR-71 Blackbird high-speed reconnaissance aircraft. At $68,000 apiece, it was excellent business for the plant, and Frank and Arlen turned out 37 units, work that continued for Arlen after the plant moved to Bakersfield and, in fact, until the Blackbird program was discontinued in 1991.

far right
Twelve-year-old Dean Lowe, in the first Kurtis quarter-midget, at a quarter-midget race in Wichita, Kansas, in 1957. The car was built in November 1955. A close examination will show that it was modeled after the 1955 Kurtis KK500D Indianapolis cars, built that spring. Young Lowe won the first quarter-midget championship at Blakely Stadium in Phoenix, Arizona, in 1957. *Dean Lowe collection*

Frank Kurtis' shop built more than three dozen of these starter carts for Air Force Lockheed F-100 Super Sabre jet fighters in the late 1950s. Later, his son, Arlen, would build starters for the super-secret SR-71 "Blackbird" reconnaissance plane. *Frank Kurtis collection*

After he retired from building racing cars, Frank Kurtis built a number of experimental vehicles for the army and the air force. This was an army amphibious vehicle called TerraStar, a Lockheed Aircraft Corporation project, made in 1967. It was designed to run equally well in water, on hard-surfaced roads, or on soft soil. *Frank Kurtis collection*

BUZZ BOMB
93
KURTIS

trailing linkages. He added 12-inch Chrysler brakes using Lincoln drums. The rear axle was also Lincoln, from the earlier Carrera car, with a limited-slip differential.

The engine he installed was a 392-cubic-inch Chrysler Hemi V-8, modified chiefly for durability with battery ignition, a modified cam, and Hilborn fuel injection. Miller estimated the engine was capable of putting out 400 horsepower, which he thought enough for a top speed of 180 miles an hour with the 1,700-pound car, while remaining nearly bulletproof. They used a four-speed Jaguar transmission mounted to the Chrysler with a Zidar adapter.

Jack Sutton, an Englishman and builder of the bodies for several land speed record cars, was then working for Kurtis building sports car sheet metal. Sutton designed an aluminum body somewhat in the style of the contemporary Maserati with touches of Osca. Sutton welded the body to a light subframe, making it a permanent, stressed, part of the structure, unlike Kurtis' Dzus-fastened panels that were easy to remove for servicing. Sutton put two headrest-covered roll bars on the tail of the car, a design derived from the Mercedes 300 SLR.

In January 1957, Sutton trailered the chassis with its new body to Miller's shop in Whittier, California. Soon after it arrived a kerosene heater overflowed and set the garage on fire. The flames melted much of Sutton's beautiful body, and destroyed the upholstery, tires, brake lines, and other flammable items. The car went back to Glendale, where Sutton rebuilt the body while Miller and his crew restored the engine and running gear.

Despite their hopes, the Carrera was not to run in 1957. Without the Mexican race as a challenge, Miller decided to run in the 24th Mille Miglia—thousand-mile—roadrace in Italy. For that effort they were able to attract sponsorship from Gerry Grant of Grant Piston Rings, Mobil Oil, and Firestone, as well as *Hot Rod* magazine, and the National Hot Rod Association. Cash donations paid the cost of shipping Miller, codriver Doug Harrison, the car, and their tools to Italy aboard the liner SS *United States*.

They arrived at Le Havre on April 23 and were met by journalists Jesse Alexander and Pete Coltrin in Coltrin's Olds 88 sedan, and proceeded at high speed across Europe. After adventures crossing Switzerland's Simplon Pass in a snowstorm, the entourage reached the starting point with just 12 days to prepare and practice before the race.

The Mille Miglia began in Brescia, in northern Italy, and ran south to Rome and back up through Bologna to its starting point. The competition included drivers such as Stirling Moss, Piero Taruffi, Jean Behra, Wolfgang von Trips and a long list of other experienced professionals in factory-prepared cars. Miller and Harrison had to drive the course in advance and make notes on each turn and intersection and have a canvas top built to meet an unexpected requirement from the Federation Internationale de L'Automobile. (A "convertible" top for the Kurtis was hastily fabricated.) Volunteers were set up at strategic spots with fuel and tires. Film crews were stationed along the route. A host of overseas Americans, many of them servicemen stationed in Germany, helped Miller and Harrison get the car to the starting line.

Three hundred and one starters left Brescia at one-minute intervals. Though their race started auspiciously and they were able to stay with Piero Taruffi in a Ferrari, Miller was about 300 miles down the coast when a brake drum broke. Later an exhaust manifold came loose from the pounding of the rough Italian roads. They were not alone, Moss' Maserati lost its brakes, the Mercedes team lost all their radiator water, and other cars rapidly fell by the wayside.

After doing some roadside repairs, Miller and Harrison motored, slowly, back to Brescia. Not only was their Mille Miglia over, but the famed road race had been run, in earnest, for the last time. It was revived as a vintage event in the 1990s, but never again would be run as an all-out race.

Back in the United States, Miller took *El Caballo* to Bonneville that summer and turned 177.420 miles per hour in the D Modified sports class. The Chrysler burned a piston, and they were not able to make a record run. The following year, 1958, Miller entered the car in the first Los Angeles Times sports Grand Prix at Riverside, but crashed in practice and was not able to make the race.

THE SECOND CITY OF INDUSTRY AND THE KURTIS *AGUILA*

Long after production of the 500KKs, at least two unfinished frames hung against the ceiling of the shop on Colorado Street. There are several tales about the old, unfinished, race car frames in Kurtis' shop and reports that he converted them into the Herb Stelter, Sam Parriott, and Jack Lufkin cars of 1962–1964. Lufkin is positive his car started life as an Indy frame, but others say the three were leftover Kurtis 500 kit frames for sports cars like that of Bill Stroppe's.

A one-time motorcycle rider, Herb Stelter of Houston, Texas, wanted to go into four-wheel vehicle racing.

Sam Parriott in his Kurtis streamliner at the starting line at the 1964 National Hot Rod Association Winternationals at Pomona, California. *Greg Sharp Collection*

He watched the 1961 Indianapolis 500 and wondered if he could get a retired Indy chassis and convert it into a killer of a sports car. He suggested the idea to his mechanic, Dale Burt, who immediately thought of Frank Kurtis. Burt and Kurtis put their heads together and decided they could build a dual purpose car, one that could be raced as a sports car, with fenders and other road equipment, or as a formula car, with open wheels. This was common practice in the 1950s. The Talbot Lagos, Giaurs, and a few other makes with removable fenders could race either way. SCCA had proposed a "formula 366" for cars based on stock passenger car blocks of no more than 366 cubic inches—5.9 liters— and Stelter, Burt, and Kurtis agreed that is what they would build.

In 1962 Stelter paid Kurtis $12,000 to build a car around one of those frames. Serial numbered 62-S-1, it was one of the last three racing cars Kurtis would build. *Aguila*, Spanish for eagle, was the name Stelter gave to the car.

Barry Knowlton, Sylvester Fredericks, Harry Pitford, and Jim Burrell put together a sleek aluminum body for the Stelter car, one unlike any of the other Kurtis sports cars. The following year Sam Parriott, who had earlier owned a Kurtis 500S, found out about another unfinished frame and had it built into a new sports drag racer also called the *City of Industry Special*.

Kurtis drew up a swoopy, streamlined aluminum body with a scheme similar to that of the 1955 Sumar, by which the front and rear fenders on each side could be removed using a few Dzus fasteners. Kurtis had built the 1955 Sumar Streamliner in that manner, and it raced at Indianapolis without its fenders because driver

George Shelley in the restored Kurtis *Aguila* racer. *George Shelley collection*

Jimmy Daywalt was uncomfortable with covered wheels. With its fenders on, the Stelter car was a super-modern, for its day, front-engine, rear-drive sports racer. Without the fenders it looked for all the world like an Indianapolis car with dual outside exhausts. Its weight, at 1,650 pounds, was about that of a then-current Kurtis Indianapolis car.

Burt installed a 327-cubic-inch Chevy V-8 with headers and a three-carburetor setup, painted it red, white, and blue and lettered its Spanish name on the cowl.

Stelter and Burt went through the usual throes of sorting out a race car. With little or no four-wheel experience, the driver was not able to be much help to the mechanic. Stelter was teaching himself to handle a powerful sports racer at the same time he was trying to tell Burt what changes were needed. As anyone who has raced knows, that was an almost impossible task. Stelter crashed or the engine had problems in virtually every outing. What was more, Formula 366 did not last, and SCCA dropped it, forcing the *Aguila* to compete in faster company.

Stelter ran the car more than a dozen times in Southwestern and Midwestern SCCA races in 1964 and 1965 without notable success. At the time mid- and rear-engine designs were taking over sports car racing and even the Reventlow Scarabs and Jim Hall's Chaparral front-engine cars were no longer competitive. Stelter eventually parked the *Aguila* for good in 1964. It sat in Burt's Houston warehouse for 20 years, was bought by a collector in Colorado, and sat for 10 more years until George Shelley discovered it, bought it, and restored it in the early 1990s. According to Shelley, it had been repaired, inaccurately, when he found the car. A new aluminum nose and the special, removable fenders were reconstructed. Shelley has run the car very well in recent vintage events.

The *Aguila* was never a road racing success, as of course some race cars never are. (Twelve of Kurtis' Indi-

The Stelter sports car, as it rolled out of the Kurtis shop in 1962. Frank Kurtis is standing behind the car. Kurtis wrote on the back of this photo, "The last of the sports cars." *Arlen Kurtis collection*

anapolis cars never qualified for the race.) But despite its lack of a winning record, it stands as a monument and memorial to its builder.

SAM PARRIOTT STREAMLINED SPORTS CAR

For several years Sam Parriott had raced a Kurtis 500S in local drag events, but in 1962 he collaborated with Kurtis in designing a more modern sports drag racer (Serial Number S-D-6-63) that he called the *City of Industry Special II*. The car, sponsored by Ford's Autolite Division, had a Roots-blown 427 Ford V-8 engine in it and was very successful in the quarter-miles. Kurtis built a dramatic aluminum body with pointed pontoon fenders, a shape much like Fred Stettler's *Aguila* sports car although the Parriott car was never planned to run without its fenders like the *Aguila*.

The cockpit of the completed Stelter Kurtis *Aguila*, showing the Kurtis tag with the serial number 62-S1. *George Shelley collection*

The Stelter Kurtis with the removable fenders shown. The car could be run as an SCCA "Formula 366" car without fenders, or as a modified sports car with the fenders attached. The fenders on the car were closely analogous to the 1955 *Sumar Special* Number 48 that ran at Indianapolis without its streamlined fenders, after driver Jimmy Daywalt found it difficult to drive without being able to see his wheels. A year after building the Stelter car, Kurtis built Sam Parriott's City of Industry sports car with an almost identical body, though Parriott apparently never drove it without the fenders. *George Shelley collection*

Parriott, who was the National Hot Rod Association's oldest competitor, drove his sleek Kurtis in dozens of drag events in Southern California. He set a class record of 148.51 miles per hour, running a 10.55 seconds elapsed time, at the Hot Rod Nationals in Indianapolis in 1963. He took the Winternationals title at Pomona at 144.69 miles per hour and won the AA/MSP class at Indianapolis in 1964 with his Kurtis, going 147.54 miles per hour with a 10.41 second elapsed time in the quarter-mile. His final drag championship came at the 1965 Pomona Winternationals. The car was, at the time, Ford's first foray into organized drag racing.

Parriott died at 82 in 1983. *The City of Industry* car has been displayed for a number of years at the Sheraton Hotel there.

LAST ED WALSH CAR

Ed Walsh commissioned Kurtis to build him one last car in 1962 to carry a rear-mounted three-cylinder, two-stroke Saab engine for H-Modified SCCA competition. It had independent suspension using coil springs, a most un-Kurtis-like design. Jim Broadwell built the fiberglass body, and Walsh did well with it in sports car racing.

LUFKIN CARS

Jack Lufkin, a member of Ak Miller's staff, was both a Bonneville racer and a drag racer. In 1962 he put together a racing Chevrolet Corvette, using a junked 1960 Corvette chassis that he and his partner, Denny Jones, hauled out of a wrecker's yard. They lightened the frame as far as they could and had Frank Kurtis build a roll bar and a front suspension for it using a solid axle and torsion bars. Kurtis installed aluminum racing bucket seats and Jim Burrell, working at Kurtis, helped Lufkin install a channeled fiberglass replica Corvette body. According to Kurtis records it bore Serial Number D-4-62. Jones painted the car a flaming red orange and waxed and buffed it until it fairly gleamed; in fact, it was known in Southern California almost as much for its finish as for its speed.

Lufkin installed a 301-cubic-inch 1964 Chevrolet V-8 with 13:1 compression using Hilborn fuel injection. He and Jones took it to Bonneville for Speed Week during the summer of 1963, and turned in record runs averaging 193.440 miles per hour. Back in Southern California, Lufkin took it to the National Hot Rod Association drag races at Pomona and turned 131.950 miles per hour in the quarter-mile in the B Modified Sports class. He continued to run the car in both drag races and at Bonneville and at El Mirage Dry Lake for two years.

In 1964 Lufkin had Kurtis build him an even faster, slicker car to run in the Bonneville modified sports car class. Using an unfinished Indianapolis frame that was gathering cobwebs in the back of the Kurtis shop, Kurtis' welder, Harry Pitford, constructed a two-seat sports car streamliner. Kurtis also built a unique all-aluminum body for the car. (Though it bore some slight resemblance to Sam Parriott's *City of Industry* Kurtis modified sports drag car, the only real connection was that Bill Edwards, who built the Parriott car, also worked on Lufkin's *Streamliner*.)

In 1965, powered by a 371-cubic-inch Chevy and with co-owner Danny Eames at the wheel, the Lufkin car qualified at Bonneville over 184, though valvetrain problems slowed their record attempt into the 170s. They ran a 255-cubic-inch Ford in the car in 1966. The *Streamliner*, running in the C Modified Sports class, competed at Bonneville for a decade, eventually topping 300 miles per hour and, in 1973, setting the top speed of the meet. Over 10 years it carried many different engines and set records in every displacement class for modified sports cars. Lufkin's accomplishments with the two Kurtis cars earned him recognition in the Dry Lakes Hall of Fame in Buellton, California.

Lufkin's Bonneville car was not only the last Indy car frame Frank Kurtis would build, but the Lufkin *Streamliner* became the last full-sized race car of any kind to roll out of the doors of Kurtis-Kraft. It was the end of a line of racing cars that had begun 30 years earlier on the oiled banks of Legion Ascot Speedway, that raced on a thousand midget tracks, on the mile dirt tracks of the Championship Trail, and passed with glory down the bricks of Indianapolis.

Jack Lufkin's Streamlined Kurtis Bonneville sports car. Lufkin drove this car for 10 years on the salt and it held records in every class in which it competed, with a top speed of 245 miles per hour in 1968. It was powered by a 396-cubic-inch, unblown Chevrolet V-8 equipped with an Iskenderian cam. This car was probably the fastest Kurtis ever built. It ran in the single-engine, unsupercharged gas modified sports car class. Lufkin was manager of the Ak Miller Garage. *Greg Sharp collection*

The Lufkin Kurtis sports streamliner was built for him by Frank Kurtis in 1964, using portions of an Indianapolis car Frank had started to build some years earlier. The design evolved over several years as different engines were used. At one time it had a bulge on the hood to cover a blower, and at another time, Lufkin used a complete canopy over the cockpit. *Greg Sharp collection*

95
PERFECT CIRCLE
SPEED

Kurtis Epilogue

CHAPTER 7

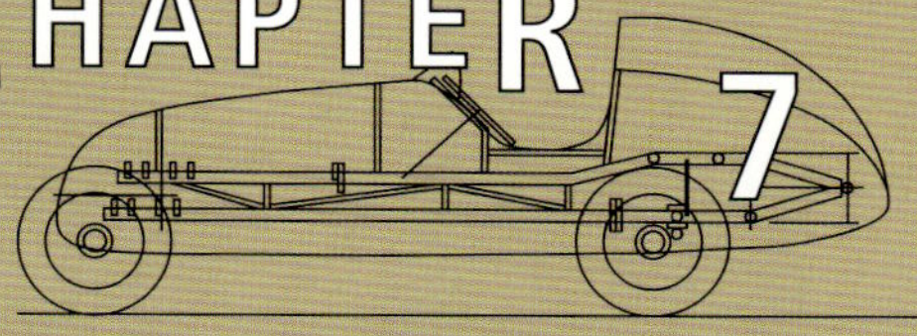

Kurtis cars ran at Indianapolis in 1964, but with A. J. Foyt's victory that year in an Offenhauser-powered Watson, the day of the roadsters was over. Never again would a Kurtis or a Watson see Victory Lane. In 1965 Tassi Vatis' *Konstant-Hot Special* carried Arnie Knepper to 18th place, the last Kurtis to turn a competitive wheel in the 500. Scotsman Jim Clark's Lotus began the rear-engine era in 1965. Front-engine cars disappeared from Indianapolis forever after Bobby Grim's Ebb Rose car crashed on the first lap in 1966. The day of the front-engine car, the roadster era, and Frank Kurtis' Indianapolis career were at an end.

The last driver of a Kurtis to win a Championship race was Van Johnson at Langhorne, Pennsylvania, on June 14, 1959, in Jake Vargo's KK4000 Number 56. By the 1970s Kurtis' racing cars were old and no longer competitive, though a few dirt track cars lingered here and there. Kurtis sports cars raced for many years, and his midgets competed until the 1990s, but the day of narrow tires and cars put together without the aid of computers was past.

Kurtis still loved racing. In 1980 he told an interviewer, "Auto racing is one sport that separates the men from the boys. No other sport can give the thrill, the pathos, and the heartaches that auto racing gives." He believed that he left racing voluntarily. "If I had wanted to, I could have been deeply involved for years," he said.

Kurtis built his last Indianapolis car in 1963, and it was not a success. His last designs for competition cars were the Stelter *Aguila* road racer in 1962 and the 1963 Parriott *City of Industry* modified sports dragster. By that time Frank Kurtis Inc. was making its money from support equipment for Lockheed and North American Aviation and prototype vehicles for the military services, in Kurtis' opinion altogether much more pleasant customers than Indianapolis race car owners.

Frank Kurtis' cars ran on the dirt tracks long after their competitive days at Indianapolis were over. Here Speed McFee readies Mike Unverdorben's *Fleet Motor Lines Special* Number 95, a KK4000, at Syracuse, New York, in 1956. Behind McFee is Charlie Musselman's Traylor Number 55. Neither car qualified that day. The Kurtis 4000s ran in AAA and USAC Championship events until 1970. Spike Gelhausen's *Midwest Manufacturing Special* Number 84, originally John McDaniel's 1951 Indianapolis entry, was the last Kurtis to compete at that level, when Tom Bigelow finished 17th at Sacramento, California, October 4, 1970. *Bruce Craig*

Arlen Kurtis moved the Kurtis-Kraft plant to Bakersfield, California, in 1977. He continued to build aircraft ground support equipment and drag racing boats. Arlen and his wife, Carol, held a number of drag racing records on the water. *Arlen Kurtis*

In 1965 Kurtis' wife, Arlene, died of liver cancer. With his daughter, Ellona, he moved from Cordova Street to a new home on Parkside Avenue in Burbank. Two years later Kurtis and his long-time friend, Edith Clemons, were married.

Despite his annoyance with the racers and the racing press, Frank Kurtis did not end his life as a bitter man. On July 4, 1968, Kurtis and Edy moved to Parker, Arizona, just below Parker Dam on the Lower Colorado River and some 100 miles south of Las Vegas, Nevada. There, on land he leased from the Indians, they moved into a double-wide trailer. Kurtis built a shop alongside the house, where he said he wanted to establish a business building recreational vehicles. He tinkered with designs for boats and even a new rear-engine Indianapolis car; he did odd jobs for the Indians on the reservation and built an amphibious camper that he drove on the reservoir above Parker Dam. He did repairs and engineering for a cotton gin that served a friend's 6,000-acre farm in Parker, and worked on restoring a succession of cars, from a Jeep to a 1964-1/2 Mustang.

In his 60s, Kurtis had retired from active work. He and Edy visited the shop in Glendale during the summer months when it was unbearably hot on the desert. He continued to correspond with owners of Kurtis cars, and he entertained visitors quite graciously.

Frank and Edy hunted and fished along the Colorado, and in 1970 they drove their camper to Alaska, a 10,000-mile trip up the gravel-paved Alcan Highway. His son, Arlen, was making a living building jet engine starters for the SR-71 and constructing fiberglass drag racing boats. Arlen and old Kurtis hand Dorsey Wire began making new Kurtis sports cars and midgets for the vintage market in the 1990s, and even as this is written, you can still buy a Kurtis race car, fresh out of the shop.

When Frank Kurtis felt the business needed him, he would return to Glendale and poke around the shop for a day or two until in May 1977, Arlen moved the business to Bakersfield. One summer Frank and Edy drove up to Crested Butte to look at the site of the old Kuretic homestead. The house he was born in had been gone for many years. In 1908, houses in Crested Butte were primitive affairs, but when they returned, the village had been restored for skiers as a quaint old mining town. Frank and Edy continued on west into Utah and poked through the little towns of Joe's Valley and Helper, where Frank had played as a child, and Schofield, where he learned metalworking in his father's blacksmith shop.

Frank's heart condition became worse in the 1980s and he was no longer able to travel. He died on February 17, 1987, and services were held in the Holy Family Catholic Church in Glendale, where the family had worshipped. He was buried in Forest Lawn Memorial Park in Hollywood Hills. When Frank Kurtis died,

Frank Kurtis was 78, living in retirement in Parker, Arizona, when John Queen and Lee Coker found and restored the 500S Kurtis sports car that Bill Stroppe drove to so many road race victories in 1953. Queen and Coker took Stroppe, seen here on the right, out to Arizona to show the car to Frank. Edy Kurtis is sitting behind the wheel of the restored car. *Edy Kurtis collection*

there passed with him an era in American racing when one man, with talent and skill but little technical training, could design a race car that could win at the highest level of the sport.

As with many men, the sting of his defeats faded and the glow of his victories grew with time. Collectors and restorers came to appreciate his racing cars. Cars he had sold for a few thousand dollars came to be worth tens or hundreds of thousands. His Indianapolis cars, his midgets, and his sports cars grew to become legends. Frank Kurtis himself came to be a legend among those who built, drove, and remembered racing in America.

This is the grille Frank Kurtis made in 1933 for Johnny Pasco's Model T roadster. It was the first race car Kurtis worked on, and it showed up at a Southern California flea market in 1980. *Edy Kurtis collection*

Appendix

Driver	Fin Pos	Car Name	Entrant	Chassis	Engine	Start Pos	Laps	Time/Speed/Out	Purse
1941									
Indianapolis, IN — 500 Mile — 30 May									
Sam Hanks	33	Tom Joyce 7-Up No.28	Ed Walsh	Kurtis	Offy		0	Wrecked after qual	500
1946									
Indianapolis, IN — 500 Mile — 30 May									
Mel Hansen	11	Offenhauser No.41	Ross Page	Kurtis	DurySc	27	143	Broken crankshaft	1,700
Ralph Hepburn	14	Novi Governor No.2	W. C. Winfield	Kurtis FD	NoviSc	19	121	Engine, stalled	7,025
George Connor	24	Walsh No.38	Ed Walsh	Kurtis	Offy	30	38	Piston	800
Langhorne, PA — 100 Mile — 30 Jun									
George Connor	9	Walsh No.38	Ed Walsh	Kurtis	Offy	4	55	Magneto	372
Atlanta, GA — 100 Mile — 2 Sep									
George Connor	1	Walsh No.38	Ed Walsh	Kurtis	Offy	8	98	No time	2,964
Indianapolis, IN — 100 Mile — 15 Sep									
George Connor	7	Walsh No.38	Ed Walsh	Kurtis	Offy	5	98	Flagged	557
Milwaukee, WI — 100 Mile — 22 Sep									
George Connor	10	Walsh No.38	Ed Walsh	Kurtis	Offy	9	46	Bad axle	519
Goshen, NY — 100 Mile — 6 Oct									
George Connor	9	Walsh No.38	Ed Walsh	Kurtis	Offy	7	55	Lost oil pressure	471
1947									
Indianapolis, IN — 500 Mile — 30 May									
Herb Ardinger	4	Novi Governor No.54	W. C. Winfield	Kurtis FD	NoviSc	4	69	4:24:32.52 113.404	6,825
Rex Mays	6	Bowes Seal Fast No.9	Bowes Racing, Inc	Kurtis	WinfSc	20	200	4:30:84.50 111.056	4,000
Fred Agabashian	9	Ross Page No.41	Ross Page	Kurtis	DurySc	23	191	Flagged	2,905
Cliff Bergere	21	Novi Governor No.18	W. C. Winfield	Kurtis FD	NoviSc	2	62	Piston	3,815
George Connor	26	Walsh No.14	Ed Walsh	Kurtis	Offy	13	32	Fuel tank leak	1,465
Cliff Bergere	98	Novi Governor No.54	W. C. Winfield	Kurtis FD	NoviSc	131		Rel Ardinger 70-200	0
Milwaukee, WI — 100 Mile — 8 Jun									
George Connor	4	Walsh No.14	Ed Walsh	Kurtis	Offy	3	100	Finished	1,699
Rex Mays	99	Bowes Seal Fast No.9	Bowes Racing, Inc	Kurtis	Winfld	0		Out in practice	0
Langhorne, PA — 100 Mile — 22 Jun									
George Connor	15	Walsh No.14	Ed Walsh	Kurtis	Offy	5	49	Engine	288
Rex Mays	99	Bowes Seal Fast No.9	Bowes Racing, Inc	Kurtis	Winfld	0		DNQ, brkn coolng web	0
Atlanta, GA — 100 Mile — 4 Jul									
Rex Mays	12	Bowes Seal Fast No.9	Bowes Racing, Inc	Kurtis	Winfld	1	0		224
George Connor	14	Walsh No.14	Ed Walsh	Kurtis	Offy	9	0	Out	168
Bainbridge, OH — 100 Mile — 13 Jul									
George Connor	5	Walsh No.14	Ed Walsh	Kurtis	Offy	4	89	Flagged - rain	477
Rex Mays	15	Bowes Seal Fast No.9	Bowes Racing, Inc	Kurtis	Winfld	12	4	Burned clutch	158
Milwaukee, WI — 100 Mile — 27 Jul									
George Connor	5	Walsh No.14	Ed Walsh	Kurtis	Offy	10	98	Flagged	801
Rex Mays	10	Bowes Seal Fast No.9	Bowes Racing, Inc	Kurtis	Winfld	8	62	Flagged	550
Goshen, NY — 100 Mile — 17 Aug									
George Connor	5	Walsh No.14	Ed Walsh	Kurtis	Offy	14	97	Flagged	398
Rex Mays	99	Bowes Seal Fast No.9	Bowes Racing, Inc	Kurtis	Winfld	0		Excused, not ready	0
Milwaukee, WI — 100 Mile — 24 Aug									
George Connor	3	Walsh No.14	Ed Walsh	Kurtis	Offy	6	100	Finished	1,341
Rex Mays	16	Bowes Seal Fast No.9	Bowes Racing, Inc	Kurtis	Winfld	9	17	Engine trouble	227
Springfield, IL — 100 Mile — 28 Sep									
Rex Mays	99	Bowes Seal Fast No.9	Bowes Racing, Inc	Kurtis	Winfld	0		Did not qualify	0
1948									
Arlington, TX — 100 Mile — 25 Apr									
Les Anderson	11	Kennedy Tank No.34	Les Anderson	Kurtis	Offy	11	50	Flagged	258
Billy DeVore	14	Pat Clancy No.19	Pat Clancy	Kurtis 6Wh	Offy	10	16	Out	203

Driver	Fin Pos	Car Name	Entrant	Chassis	Engine	Start Pos	Laps	Time/Speed/Out	Purse
Rex Mays	16	Bowes Seal Fast No.5	Bowes Racing, Inc.	Kurtis	WinfSc	3	8	Out	180
Walt Brown	99	Kurtis Kraft No.7	Ed Walsh	Kurtis	Offy		0	Did not qualify	44
Indianapolis, IN — 500 Mile — 31 May									
Duke Nalon	3	Novi Grooved Piston No.54	W. C. Winfield	Kurtis FD	NoviSc	11	200	4:14:09.78 118.034	15,675
Mack Hellings	5	Don Lee No.35	Don Lee Division	Kurtis 2000	Offy	21	200	4:24:38.52 113.361	7,675
Hal Cole	6	City of Tacoma No.63	R. Taylor	Kurtis 2000	Offy	14	200	4:28:50.86 111.587	5,425
Tom Hinnershitz	9	Kurtis Kraft No.7	Ed Walsh	Kurtis	Offy	23	198	Flagged	4,270
Billy DeVore	12	Pat Clancy No.19	Pat Clancy	Kurtis 6Wh	Offy	20	190	Flagged	2,930
Johnny Mantz	13	Agajanian No.98	J.C. Agajanian	Kurtis 2000	Offy	8	185	Flagged	2,230
Rex Mays	19	Bowes Seal Fast No.5	Bowes Racing, Inc	Kurtis	WinfSc	1	129	Fuel tank leak	5,775
Fred Agabashian	23	Ross Page No.26	Ross Page	Kurtis	DurySc	32	58	Broken oil line	1,580
Les Anderson	24	Kennedy Tank No.34	Les Anderson	Kurtis	Offy	9	58	Broken gear	1,550
Emil Andres	31	Tuffy's Offy No.8	C. George Tuffanelli	Kurtis 2000	Offy	16	11	Steering	1,340
Ralph Hepburn	99	Novi Grooved Piston No.12	W. C. Winfield	Kurtis 2000	NoviSc	0		Wrecked practice	0
Andy Granatelli	99	Grancor V8 No.39	J. & A. Granatelli	Kurtis 2000	Mercry	0		Did not qualify	0
Andy Granatelli	99	Grancor-Werner No.59	Andy Granatelli	Kurtis 2000	Mercry	0		Wrecked practice	0
Cliff Bergere	99	Grancor-Werner No.85	Anthony Granatelli	Kurtis 2000	Offy		0	Did not qualify	0
Manuel Ayulo	99	Weidel No.88	Conrad Weidel	Kurtis	Mercry	0		Did not qualify	0
Milwaukee, WI — 100 Mile — 6 Jun									
Emil Andres	1	Tuffy's Offy No.8	C. George Tuffanelli	Kurtis 2000	Offy	3	100	1:10:19.43 85.319	5,250
Mack Hellings	2	Don Lee No.35	Mal Ord	Kurtis 2000	Offy	8	100	Finished	3,780
Hal Cole	6	City of Tacoma No.63	R. Taylor	Kurtis 2000	Offy	9	100	Finished	1,050
Johnny Mantz	12	Agajanian No.98	J.C. Agajanian	Kurtis 2000	Offy	1	89	Broken piston	525
Walt Brown	13	Kurtis Kraft No.7	Ed Walsh	Kurtis	Offy	10	81	Ignition trouble	420
Rex Mays	17	Bowes Seal Fast No.5	Bowes Racing, Inc.	Kurtis	WinfSc	11	6	Wrecked	210
Les Anderson	99	Kennedy Tank No.34	Les Anderson	Kurtis	Offy		0	DNS, too slow	0
Manuel Ayulo	99	Bob Estes No.88	Conrad Weidel	Kurtis	Offy		0	DNS, too slow	0
Langhorne, PA — 100 Mile — 20 Jun									
Walt Brown	1	Kurtis Kraft No.7	Ed Walsh	Kurtis	Offy	11	100	1:06:55.66 89.628	2,585
Mack Hellings	2	Don Lee No.35	Mal Ord	Kurtis 2000	Offy	9	100	Finished	1,847
Emil Andres	3	Pritchard No.8	Charles Pritchard	Kurtis	Offy	4	100	Finished	1,103
Johnny Mantz	13	Agajanian No.98	J.C. Agajanian	Kurtis 2000	Offy	5	43	Out	211
Rex Mays	16	Bowes Seal Fast No.5	Bowes Racing, Inc.	Kurtis	WinfSc	6	21	Engine trouble	187
Hal Cole	99	City of Tacoma No.63	R. Taylor	Kurtis 2000	Offy		0	DNS, too slow	0
Milwaukee, WI — 100 Mile — 15 Aug									
Johnny Mantz	1	Agajanian No.98	J.C. Agajanian	Kurtis 2000	Offy	3	100	1:10:19.08 85.326	3,978
Emil Andres	2	Pritchard No.8	Charles Pritchard	Kurtis	Offy	9	100	Finished	2,843
Rex Mays	4	Bowes Seal Fast No.5	Bowes Racing, Inc.	Kurtis	WinfSc	5	100	Finished	1,145
Walt Brown	13	Kurtis Kraft No.7	Ed Walsh	Kurtis	Offy	10	80	Out	345
Mack Hellings	14	Grancor No.59	Andy Granatelli	Kurtis 2000	Offy	8	78	Out	345
Springfield, IL — 100 Mile — 21 Aug									
Johnnie Parsons	2	Kurtis Kraft No.7	Ed Walsh	Kurtis	Offy	14	100	Finished	1,890
Rex Mays	4	Bowes Seal Fast No.5	Bowes Racing, Inc	Kurtis	WinfSc	7	100	Finished	735
Johnny Mantz	9	Agajanian No.98	J.C. Agajanian	Kurtis 2000	Offy	6	97	Flagged	315
Hal Robson	10	Ross Page No.26	Ross Page	Kurtis		16	90	Out	315
Emil Andres	18	Pritchard No.8	Charles Pritchard	Kurtis	Offy	8	2	Wrecked with No.12	102
Mack Hellings	19	Grancor No.59	Andy Granatelli	Kurtis 2000	Offy	12	2	Wrecked with No.12	102
Milwaukee, WI — 200 Mile — 29 Aug									
Johnnie Parsons	2	Kurtis Kraft No.7	Ed Walsh	Kurtis	Offy	12	200	Finished	3,685
Rex Mays	7	Bowes Seal Fast No.5	Bowes Racing, Inc.	Kurtis	WinfSc	4	182	Magneto	973
Johnny Mantz	13	Agajanian No.98	J.C. Agajanian	Kurtis 2000	Offy	9	0	Out	603
Hal Robson	14	Ross Page No.26	Ross Page	Kurtis		14	0	Out	591
Emil Andres	16	Pritchard No.8	Charles Pritchard	Kurtis	Offy	7	0	Out	530
DuQuoin, IL — 100 Mile — 4 Sep									
Hal Robson	6	Ross Page No.26	Ross Page	Kurtis		12	100	Finished	500
Johnnie Parsons	11	Kurtis Kraft No.7	Ed Walsh	Kurtis	Offy	8	87	Out	250
Rex Mays	17	Bowes Seal Fast No.5	Bowes Racing, Inc.	Kurtis	WinfSc	1	5	Brkn connecting rod	100
Atlanta, GA — 100 Mile — 6 Sep									
Hal Robson	4	Ross Page No.26	Ross Page	Kurtis		7	100	Finished	651
Rex Mays	10	Bowes Seal Fast No.5	Bowes Racing, Inc	Kurtis	WinfSc	2	28	Broken fuel tank	279

Kurtis-Kraft

Driver	Fin Pos	Car Name	Entrant	Chassis	Engine	Start Pos	Laps	Time/Speed/Out	Purse
Springfield, IL — 100 Mile — 19 Sep									
Emil Andres	2	Pritchard No.8	Charles Pritchard	Kurtis	Offy	5	100	Finished	1,520
Johnny Mantz	9	Agajanian No.98	J.C. Agajanian	Kurtis 2000	Offy	4	100	Finished	253
Johnnie Parsons	13	Kurtis Kraft No.7	Ed Walsh	Kurtis	Offy	13	50	Brkn steering arm	169
Rex Mays	14	Bowes Seal Fast No.5	Bowes Racing, Inc	Kurtis	WinfSc	6	50	Broken fuel tank	169
Hal Robson	99	Ross Page No.26	Ross Page	Kurtis			0	Did not start	0
DuQuoin, IL — 100 Mile — 10 Oct									
Johnnie Parsons	1	Kurtis Kraft No.7	Ed Walsh	Kurtis	Offy	7	100	1:11:47.70 83.572	1,875
Hal Cole	5	Grancor No.59	Andy Granatelli	Kurtis 2000	Offy	8	100	Finished	450
Emil Andres	13	Pritchard No.8	Charles Pritchard	Kurtis	Offy	9	40	Out	150
Johnny Mantz	16	Agajanian No.98	J.C. Agajanian	Kurtis 2000	Offy	6	2	Wrecked	150
Rex Mays	18	Bowes Seal Fast No.5	Bowes Racing, Inc	Kurtis	WinfSc	1	1	Out	75
1949									
Arlington, TX — 100 Mile — 24 Apr									
Johnnie Parsons	1	Kurtis-Kraft No.12	Kurtis-Kraft, Inc.	Kurtis	Offy	6	95	Finished	2,143
Rex Mays	2	Wolfe No.15	Ervin Wolfe	Kurtis 2000	Offy	1	0		1,643
Johnny Mantz	3	Agajanian No.98	J.C. Agajanian	Kurtis 2000	Offy	7	0		857
Indianapolis, IN — 500 Mile — 30 May									
Johnnie Parsons	2	Kurtis Kraft No.12	Kurtis-Kraft, Inc.	Kurtis	Offy	12	200	4:10:26.97 119.785	18,250
Joie Chitwood	5	Wolfe No.77	Ervin Wolfe	Kurtis 2000	Offy	16	200	4:12:36.97 118.757	6,950
Johnny Mantz	7	Agajanian No.98	J.C. Agajanian	Kurtis 2000	Offy	9	200	4:16:06.01 117.601	4,690
Paul Russo	8	Tuffy's Offy No.19	Charles Pritchard	Kurtis 2000	Offy	19	200	4:28:11.28 111.862	4,940
Emil Andres	9	Tuffy's Offy No.9	Charles Pritchard	Kurtis 2000	Offy	31	18	Flagged	4,420
Mack Hellings	16	Don Lee No.8	Don Lee Motors Corp.	Kurtis 2000	Offy	14	172	Valve	2,570
Hal Cole	19	Grancor No.14	Grancor Auto Specialists	Kurtis 2000	Offy	11	117	Broken rod bearing	1,975
Jackie Holmes	22	Pat Clancy No.57	Pat Clancy	Kurtis 6Wh	Offy	17	65	Broken drive shaft	2,410
Bayliss Levrett	24	Wynn's Fricting Proofing No.69	Bayliss Levrett/E. Morris	Kurtis 2000	Offy	28	52	Lost drain plug	2,375
Rex Mays	25	Novi Mobil No.5	Lou Welch	Kurtis FD	NoviSc	2	48	Magneto	3,470
Jack McGrath	26	City of Tacoma No.33	Leo Dobry	Kurtis 2000	Offy	3	39	Broken oil pump	2,115
Duke Nalon	29	Novi Mobil No.54	Lou Welch	Kurtis FD	NoviSc	1	23	Wrecked NE, on fire	5,650
Sam Hanks	30	Love Machine & Tool No.18	Milt Marion	Kurtis 2000	Offy	23	20	Oil leak	1,645
Walt Brown	98	Tuffy's Offy No.9	Charles Pritchard	Kurtis 2000	Offy		179	Rel Andres, 19-197	0
Doc Williams	99	Sarafoff No.65	Tom Sarafoff	Kurtis 2000	Offy		0	DNS, too slow	0
Ed Haddad	99	Page No.47	Ross Page	Kurtis	DurySc	0		Did not qualify	0
Milwaukee, WI — 100 Mile — 5 Jun									
Emil Andres	2	Tuffy's Offy No.9	Charles Pritchard	Kurtis 2000	Offy	6	100	Finished	3,708
Mack Hellings	3	Don Lee No.8	Don Lee Motors Corp.	Kurtis 2000	Offy	18	100	Finished	2,060
Johnny Mantz	5	Agajanian No.98	J.C. Agajanian	Kurtis 2000	Offy	5	100	Finished	1,236
Paul Russo	6	Tuffy's Offy No.19	Charles Pritchard	Kurtis 2000	Offy	8	100	Finished	1,030
Rex Mays	7	Wolfe No.77	Ervin Wolfe	Kurtis 2000	Offy	1	100	Finished	824
Jack McGrath	8	City of Tacoma No.33	Leo Dobry	Kurtis 2000	Offy	11	100	Finished	824
Walt Brown	10	Love Machine & Tool No.18	Milt Marion	Kurtis 2000	Offy	7	100	Finished	618
Johnnie Parsons	13	Kurtis-Kraft No.12	Kurtis-Kraft, Inc.	Kurtis	Offy	17	92	Flagged	412
Bayliss Levrett	14	Offenhauser No.69	Bayliss Levrett/E. Morris	Kurtis 2000	Offy	4	78	Out	412
Hal Cole	99	Grancor No.14	Grancor Auto Specialists	Kurtis 2000	Offy	0		DNS, too slow	0
Trenton, NJ — 100 Mile — 19 Jun									
Walt Brown	3	Milt Marion No.18	Milt Marion	Kurtis 2000	Offy	14	100	Finished	859
Emil Andres	4	Tuffy's Offy No.9	Charles Pritchard	Kurtis 2000	Offy	13	100	Finished	601
Duke Dinsmore	5	Wynn's Friction Proofing No.69	Bayliss Levrett/E. Morris	Kurtis 2000	Offy	10	100	Finished	515
Paul Russo	8	Tuffy's Offy No.19	Charles Pritchard	Kurtis 2000	Offy	9	0		343
Rex Mays	10	Wolfe No.77	Ervin Wolfe	Kurtis 2000	Offy	1	0		257
Johnnie Parsons	11	Kurtis-Kraft No.12	Kurtis-Kraft, Inc.	Kurtis	Offy	8	0		214
Johnny Mantz	12	Agajanian No.98	J.C. Agajanian	Kurtis 2000	Offy	7	0		214
Jack McGrath	14	City of Tacoma No.33	Leo Dobry	Kurtis 2000	Offy	6	0		171
Hal Cole	17	Cole No.14	Grancor Auto Specialists	Kurtis 2000	Offy	17	0		85
Springfield, IL — 100 Mile — 20 Aug									
Rex Mays	2	Wolfe No.15	Ervin Wolfe	Kurtis 2000	Offy	2	100	Finished	1,800
Johnnie Parsons	3	Kurtis-Kraft No.12	Kurtis-Kraft, Inc.	Kurtis	Offy	16	100	Finished	1,000
Walt Brown	7	Tuffy's Offy No.9	Charles Pritchard	Kurtis 2000	Offy	7	0		400
Johnny Mantz	8	Agajanian No.98	J.C. Agajanian	Kurtis 2000	Offy	17	0		400
Paul Russo	10	Tuffy's Offy No.19	Charles Pritchard	Kurtis 2000	Offy	10	0		300
Duane Carter	15	Cole No.14	Grancor Auto Specialists	Kurtis 2000	Offy	4	0		200
Bayliss Levrett	17	Wynn's Friction Proofing No.69	Bayliss Levrett/E. Morris	Kurtis 2000	Offy	14	0		100
Tony Bettenhausen	18	Meyer & Drake No.99	Murrell Belanger	Kurtis	M&D	8	0		100
Ed Haddad	99	City of Tacoma No.33	Leo Dobry	Kurtis 2000	Offy		0	Did not qualify	0
Milwaukee, WI — 200 Mile — 28 Aug									
Johnnie Parsons	1	Kurtis-Kraft No.12	Kurtis-Kraft, Inc.	Kurtis	Offy	5	200	2:19:49.95 85.818	6,500

Driver	Fin Pos	Car Name	Entrant	Chassis	Engine	Start Pos	Laps	Time/Speed/Out	Purse
Walt Brown	2	Tuffy's Offy No.9	Charles Pritchard	Kurtis 2000	Offy	14	200	Finished	3,900
Rex Mays	3	Wolfe No.15	Ervin Wolfe	Kurtis 2000	Offy	2	200	Finished	2,600
Bayliss Levrett	4	Wynn's Friction Proofing No.69	Bayliss Levrett/E. Morris	Kurtis 2000	Offy	18	97	Finished	1,820
Paul Russo	12	Tuffy's Offy No.19	Charles Pritchard	Kurtis 2000	Offy	10	184	Out	697
Tony Bettenhausen	15	Meyer & Drake No.99	Murrell Belanger	Kurtis	M&D	3	127	Out	481
Johnny Mantz	17	Agajanian No.98	J.C. Agajanian	Kurtis 2000	Offy	15	63	Out	351
Duke Dinsmore	98	Wynn's Friction Proofing No.69	Bayliss Levrett/E. Morris	Kurtis 2000	Offy		103	Rel Levrett 98-200	0
Sam Hanks	99	Wales No.14	Pete Wales	Kurtis 2000	Offy		0	DNS, too slow	0
Ed Haddad	99	City of Tacoma No.33	Leo Dobry	Kurtis 2000	Offy		0	Did not qualify	0
DuQuoin, IL — 100 Mile — 3 Sep									
Tony Bettenhausen	1	Meyer & Drake No.99	Murrell Belanger	Kurtis	OffySc	100		1:06:37.13 90.065	2,164
Walt Brown	5	Tuffy's Offy No.18	Charles Pritchard	Kurtis 2000	Offy		0		519
Paul Russo	6	Tuffy's Offy No.19	Charles Pritchard	Kurtis 2000	Offy		0		433
Johnnie Parsons	11	Kurtis-Kraft No.12	Kurtis-Kraft, Inc.	Kurtis	Offy		0		217
Rex Mays	15	Neal's Rest No.14	Pete Wales	Kurtis 2000	Offy		0		173
Pikes Peak, CO — 12 Mile — 5 Sep									
Mack Hellings	5	Don Lee No.2	Don Lee Motors Corp.	Kurtis 2000	AlfRom	1		0:16:59.50 43.857	0
Syracuse, NY — 100 Mile — 10 Sep									
Johnnie Parsons	1	Kurtis-Kraft No.12	Kurtis-Kraft, Inc.	Kurtis	Offy	10	100	1:09:36.52 86.196	3,243
Walt Brown	4	Tuffy's Offy No.18	Charles Pritchard	Kurtis 2000	Offy	4	99	Flagged	908
Duane Carter	7	Neal's Rest No.14	Pete Wales	Kurtis 2000	Offy	11	97	Flagged	519
Johnny Mantz	11	Agajanian No.98	J.C. Agajanian	Kurtis 2000	Offy	7	92	Flagged	324
Rex Mays	15	Wolfe No.15	Ervin Wolfe	Kurtis 2000	Offy	2	73	Out	259
Detroit, MI — 100 Mile — 11 Sep									
Tony Bettenhausen	1	Meyer & Drake No.99	Murrell Belanger	Kurtis	OffySc	2	100	1:13:50.00 81.264	4,875
Walt Brown	8	Tuffy's Offy No.18	Charles Pritchard	Kurtis 2000	Offy	14	72	Flagged	780
Duke Dinsmore	10	Wynn's Friction Proofing No.69	Bayliss Levrett/E. Morris	Kurtis 2000	Offy	1	62	Flagged	585
Rex Mays	11	Wolfe No.15	Ervin Wolfe	Kurtis 2000	Offy	9	59	Flagged	487
Johnnie Parsons	12	Kurtis-Kraft No.12	Kurtis-Kraft, Inc.	Kurtis	Offy	6	49	Wrecked	487
Duane Carter	14	Neal's Rest No.14	Pete Wales	Kurtis 2000	Offy	8	34	Out	390
Johnny Mantz	99	Agajanian No.98	J.C. Agajanian	Kurtis 2000	Offy		0	DNS, too slow	195
Springfield, IL — 100 Mile — 25 Sep									
Johnnie Parsons	1	Kurtis-Kraft No.12	Kurtis-Kraft, Inc.	Kurtis	Offy	17	100	1:06:08.75 91.729	1,750
Rex Mays	2	Wolfe No.15	Ervin Wolfe	Kurtis 2000	Offy	9	100	Finished	1,260
Walt Brown	5	Tuffy's Offy No.18	Charles Pritchard	Kurtis 2000	Offy	15	100	Finished	420
Tony Bettenhausen	9	Belanger No.99	Murrell Belanger	Kurtis	M&D	6	0	Flagged	210
Johnny Fedricks	14	Neal's Rest No.14	Pete Wales	Kurtis 2000	Offy	16	0	Out	140
Bayliss Levrett	16	Wynn's Friction Proofing No.69	Bayliss Levrett/E. Morris	Kurtis 2000	Offy	2	0	Out	140
Langhorne, PA — 100 Mile — 16 Oct									
Johnnie Parsons	1	Kurtis-Kraft No.12	Kurtis-Kraft, Inc.	Kurtis	Offy	4	100	1:04:56.22 92.398	1,798
Walt Brown	11	Tuffy's Offy No.18	Charles Pritchard	Kurtis 2000	Offy	2	80	Flagged	179
Sacramento, CA — 100 Mile — 30 Oct									
Fred Agabashian	1	Agajanian No.98	J.C. Agajanian	Kurtis 2000	Offy	1	100	1:11:01.05 84.486	0
Rex Mays	2	Wolfe No.15	Ervin Wolfe	Kurtis 2000	Offy	3	100	Finished	0
Johnnie Parsons	5	Kurtis-Kraft No.12	Kurtis-Kraft, Inc.	Kurtis	Offy	13	100	Finished	0
Walt Brown	6	Tuffy's Offy No.18	Charles Pritchard	Kurtis 2000	Offy	5	100	Finished	0
Mack Hellings	9	Don Lee No.8	Don Lee Motors Corp.	Kurtis 2000	Offy	17	95	Flagged	0
Paul Russo	10	Tuffy's Offy No.24	Charles Pritchard	Kurtis 2000	Offy	11	68	Flagged	0
Duke Dinsmore	14	Wynn's Friction Proofing No.69	Bayliss Levrett/E. Morris	Kurtis 2000	Offy	7	57	Overheated	0
Hal Cole	15	Neal's Rest No.14	Pete Wales	Kurtis 2000	Offy	16	23	Lost oil pressure	0
Cal Niday	18	City of Tacoma No.33	Leo Dobry	Kurtis 2000	Offy	15	5	Broken piston	0
Tony Bettenhausen	99	Belanger No.99	Murrell Belanger	Kurtis	Offy		0	Did not qualify	0
Del Mar, CA — 100 Mile — 6 Nov									
Mack Hellings	4	Don Lee Motors No.8	Don Lee Motors Corp.	Kurtis 2000	Offy	18	100	Finished	0
Paul Russo	5	Tuffy's Offy No.24	Charles Pritchard	Kurtis 2000	Offy	7	98	Flagged	0
Walt Brown	6	Tuffy's Offy No.18	Charles Pritchard	Kurtis 2000	Offy	12	97	Flagged	0
Cal Niday	7	City of Tacoma No.33	Leo Dobry	Kurtis 2000	Offy	15	97	Flagged	0
Johnny Mantz	8	Agajanian No.98	J.C. Agajanian	Kurtis 2000	Offy	11	96	Flagged	0
Hal Cole	11	Neal's Rest No.14	Pete Wales	Kurtis 2000	Offy	10	14	Flagged	0
Johnnie Parsons	15	Kurtis-Kraft No.12	Kurtis-Kraft, Inc.	Kurtis	Offy	5	36	Slipping clutch	0
Tony Bettenhausen	16	Belanger No.99	Murrell Belanger	Kurtis	M&D	3	29	Out	0
Rex Mays	18	Wolfe No.15	Ervin Wolfe	Kurtis 2000	Offy	2	14	Wrecked	0
McGee	98	Neal's Rest No.14	Pete Wales	Kurtis 2000	Offy		71	Rel Cole 15-85	0
Duke Dinsmore	99	Wynn's Friction Proofing No.69	Bayliss Levrett/E. Morris	Kurtis 2000	Offy		0	Wrecked during qual	0

Kurtis-Kraft

Driver	Fin Pos	Car Name	Entrant	Chassis	Engine	Start Pos	Laps	Time/Speed/Out	Purse
1950									
Indianapolis, IN — 500 Mile — 30 May									
Johnnie Parsons	1	Wynn's Friction Proofing No.1	Ed Walsh	Kurtis	Offy	5	138	2:46:55.97 124.002	57,458
Cecil Green	4	John Zink No.54	M.A. Walker	Kurtis 3000	Offy	12	137	2:48:45.97 121.766	10,963
Joie Chitwood	5	Wolfe No.17	Ervin Wolfe	Kurtis 2000	Offy	9	85	2:47:32.99 121.755	8,788
Walt Faulkner	7	Grant Piston Ring No.98	J.C. Agajanian	Kurtis 2000	Offy	1	135	2:47:13.55 121.094	7,663
Pat Flaherty	10	Granatelli-Sabourin No.59	Grancor Auto Specialists	Kurtis 3000	Offy	11	135	2:48:49.02 119.952	4,638
Mack Hellings	13	Tuffy's Offy No.15	Charles Pritchard	Kurtis 2000	Offy	26	132	Flagged - rain	2,968
Jack McGrath	14	Hinkle No.49	Jack B. Hinkle	Kurtis 3000	Offy	6	131	Flagged, hit NE wall	2,798
Johnny McDowell	18	Peter Wales Trucking No.62	M. Peter Wales	Kurtis 2000	Offy	33	128	Flagged - rain	2,768
Walt Brown	19	Tuffy's Offy No.4	Charles Pritchard	Kurtis 2000	Offy	20	127	Flagged - rain	2,338
Jerry Hoyt	21	Dr. Morris No.81	Ludson D. Morris, MD	Kurtis 2000	Offy	15	125	Flagged - rain	2,375
Fred Agabashian	28	Wynn's Friction Proofing No.28	Kurtis Kraft, Inc.	Kurtis 3000	OffySc	2	64	Broken oil line	2,553
Jimmy Jackson	29	Cummins Diesel No.61	Cummins Engine Co.	Kurtis 3000	CummSc	32	52	Brkn SC vibr damper	1,938
Sam Hanks	30	Merz Engineering No.37	Milton Marion	Kurtis 2000	Offy	25	42	Lost oil pressure	2,133
Duke Dinsmore	33	Brown Motor No.69	Verlin Brown	Kurtis 2000	Offy	7	10	Oil leak	1,843
Tony Bettenhausen	98	Wolfe No.17	Ervin Wolfe	Kurtis 2000	Offy		51	Rel Chitwood 86-136	0
Joe James	99	Esmeralda No.63	Leo Dobry	Kurtis 2000	Offy		0	DNS, too slow	0
Ted Duncan	99	Tuffy's Offy No.16	Charles Pritchard	Kurtis 2000	Offy		0	Did not qualify	0
Hal Cole	99	Tuffy's Offy No.16	Charles Pritchard	Kurtis 2000	Offy		0	Wrecked qualifying	0
Joel Thorne	99	Thorne Engineering No.33	Thorne Engineering Co.	Kurtis	Sparks	0		Did not qualify	0
Duke Nalon	99	Novi Mobil No.38	Winfield Engineering, Inc	Kurtis FD	NoviSc	0		Did not qualify	0
Chet Miller	99	Novi Mobil No.43	Winfield Engineering, Inc	Kurtis FD	NoviSc	0		Did not qualify	0
Bill Cantrell	99	Bowes Seal Fast No.44	Bowes Racing, Inc.	Kurtis	OffySc	0		Did not qualify	0
Bob Gregg	99	Esmeralda No.63	Leo Dobry	Kurtis 2000	Offy		0	Did not qualify	0
Marvin Burke	99	Offenhauser\Ross Page No.65	Ross Page	Kurtis	DurySc	0		Did not qualify	0
Norm Houser	99	Offenhauser\Ross Page No.65	Ross Page	Kurtis	DurySc	0		Did not qualify	0
Kenny Eaton	99	Belanger No.99	Murrell Belanger	Kurtis	OffySc	0		Did not qualify	0
Emil Andres	99	Belanger No.99	Murrell Belanger	Kurtis	OffySc	0		Did not qualify	0
Milwaukee, WI — 100 Mile — 11 Jun									
Walt Faulkner	7	Grant Piston Ring No.98	J.C. Agajanian	Kurtis 2000	Offy	17	99	Flagged	800
Fred Agabashian	8	Wynn's Friction Proofing No.28	Kurtis Kraft, Inc.	Kurtis 3000	OffySc	10	94	Flagged	800
Jack McGrath	9	Hinkle No.49	Jack B. Hinkle	Kurtis 3000	Offy	14	91	Flagged	600
Johnnie Parsons	11	Granatelli-Sabourin No.1	Grancor Auto Specialists	Kurtis 3000	Offy	13	87	Broken piston	500
Johnny McDowell	12	Peter Wales Trucking No.62	M. Peter Wales	Kurtis 2000	Offy	8	49	Engine quit	400
Mack Hellings	13	Tuffy's Offy No.15	Charles Pritchard	Kurtis 2000	Offy	11	62	Stalled	500
Duke Dinsmore	14	Brown Motor No.69	Verlin Brown	Kurtis 2000	Offy	7	52	Hit in head by rock	400
Walt Brown	15	Tuffy's Offy No.4	Charles Pritchard	Kurtis 2000	Offy	15	43	Broken rear axle	400
Emil Andres	16	Ross Page No.23	Ross Page	Kurtis	OffySc	18	43	Fouled plugs	400
Chuck Stevenson	98	Peter Wales Trucking No.62	M. Peter Wales	Kurtis 2000	Offy		32	Rel McDowell 50-81	0
Duke Nalon	99	Esmeralda No.63	Leo Dobry	Kurtis 2000	Offy		0	DNS, too slow	0
Cecil Green	99	John Zink No.54	M.A. Walker	Kurtis 3000	Offy		0	DNS, too slow	0
Duane Carter	99	Belanger No.99	Murrell Belanger	Kurtis	Offy		0	DNS, too slow	0
Jerry Hoyt	99	Dr. Morris No.81	Ludson D. Morris, MD	Kurtis 2000	Offy		0	Hit wall in practice	0
Langhorne, PA — 100 Mile — 25 Jun									
Jack McGrath	1	Hinkle No.49	Jack B. Hinkle	Kurtis 3000	Offy	5	100	1:07:47.01 88.517	1,858
Duke Dinsmore	4	Brown Motor No.69	Verlin Brown	Kurtis 2000	Offy	9	65	Finished	551
Duke Nalon	5	Chapman No.17	H.A. Chapman	Kurtis 2000	Offy	15	100	Finished	398
Fred Agabashian	6	Wynn's Friction Proofing No.28	Kurtis Kraft, Inc.	Kurtis 3000	OffySc	18	74	Finished	331
Mack Hellings	7	Tuffy's Offy No.16	Charles Pritchard	Kurtis 2000	Offy	11	99	Flagged	268
Jerry Hoyt	8	Dr. Morris No.81	Ludson D. Morris, MD	Kurtis 2000	Offy	13	99	Flagged	268
Cecil Green	9	John Zink No.54	M.A. Walker	Kurtis 3000	Offy	12	35	Flagged	268
Andy Linden	18	Ross Page No.23	Ross Page	Kurtis	OffySc	17	10	Broken piston	110
Lee Wallard	98	Wynn's Friction Proofing No.28	Kurtis Kraft, Inc.	Kurtis 3000	OffySc	26		Rel Agabash'n 51-76	0
Cliff Griffith	98	John Zink No.54	M.A. Walker	Kurtis 3000	Offy		49	Rel Green 22-70	0
Ralph Pratt	98	John Zink No.54	M.A. Walker	Kurtis 3000	Offy		14	Rel Green 85-98	0
Johnnie Parsons	98	Brown Motor No.69	Verlin Brown	Kurtis 2000	Offy		35	Rel Dinsmore 25-59	0
Johnnie Parsons	99	Grancor No.1	Grancor Auto Specialists	Kurtis 3000	Offy		0	DNS, too slow	0
Harry Turner	99	Belanger No.99	Murrell Belanger	Kurtis	Offy		0	DNS, too slow	0
Walt Brown	99	Tuffy's Offy No.4	Charles Pritchard	Kurtis 2000	Offy		0	Brkn axle, practice	0
Mike Nazaruk	99	Marion Engr No.23	Milt Marion	Kurtis 2000	Offy		0	Did not qualify	0
Buster Warke	99	Warke No.83	Buster Warke	Kurt\Warke	Offy		0	Flipped practice	0
Walt Faulkner	99	Grant Piston Ring No.98	J.C. Agajanian	Kurtis 2000	Offy		0	Brkn conn rod pract	0
Springfield, IL — 100 Mile — 19 Aug									
Mack Hellings	3	Tuffy's Offy No.15	Charles Pritchard	Kurtis 2000	Offy	18	100	Finished	1,035
Duke Dinsmore	7	Brown Motor No.69	Verlin Brown	Kurtis 2000	Offy	11	99	Flagged	460
Fred Agabashian	8	Wynn's Friction Proofing No.28	Kurtis Kraft, Inc.	Kurtis 3000	OffySc	3	99	Flagged	431
Jack McGrath	14	Hinkle No.49	Jack B. Hinkle	Kurtis 3000	Offy	9	0	Out	230
Walt Brown	16	Tuffy's Offy No.4	Charles Pritchard	Kurtis 2000	Offy	5	4	Wrecked turn 2	172
Duane Carter	17	Belanger No.99	Murrell Belanger	Kurtis	Offy	7	4	Wrecked turn 2	115
Johnnie Parsons	99	Granatelli No.1	Grancor Auto Specialists	Kurtis 3000	Offy		0	DNS, too slow	0
Neal Carter	99	Marion Engr No.23	Milt Marion	Kurtis 2000	Offy		0	DNS, too slow	0

Driver	Fin Pos	Car Name	Entrant	Chassis	Engine	Start Pos	Laps	Time/Speed/Out	Purse
Duke Nalon	99	Chapman No.17	H.A. Chapman	Kurtis 2000	Offy		0	DNS, too slow	0
Bobby Barker	99	Pritchard No.16	Charles Pritchard	Kurtis 2000	Offy		0	DNS, too slow	0
Cecil Green	99	John Zink No.54	M.A. Walker	Kurtis 3000	Offy		0	DNS, too slow	0
Walt Faulkner	99	Grant Piston Ring No.98	J.C. Agajanian	Kurtis 2000	Offy		0	DNS, too slow	0
Kenny Eaton	99	Pete Wales No.62	M. Peter Wales	Kurtis 2000	Offy		0	DNS, too slow	0
Jerry Hoyt	99	Dr. Morris No.81	Ludson D. Morris, MD	Kurtis 2000	Offy		0	Burned piston	0
Milwaukee, WI — 200 Mile — 27 Aug									
Walt Faulkner	1	Grant Piston Ring No.98	J.C. Agajanian	Kurtis 2000	Offy	8	200	2:17:26.05 87.315	6,640
Duke Dinsmore	4	Brown Motor No.69	Verlin Brown	Kurtis 4000	Offy	18	200	Finished	1,723
Duke Nalon	6	Chapman No.17	H.A. Chapman	Kurtis 2000	Offy	21	200	Finished	1,325
Fred Agabashian	8	Wynn's Friction Proofing No.28	Kurtis Kraft, Inc.	Kurtis 3000	OffySc	11	146	Flagged	994
Cecil Green	12	John Zink No.54	M.A. Walker	Kurtis 3000	Offy	12	186	Flagged	662
Neal Carter	17	Marion Engr No.23	Milt Marion	Kurtis 2000	Offy	9	140	Main bearing	265
Chuck Stevenson	20	Belanger No.99	Murrell Belanger	Kurtis	Offy	1	37	Wrecked north turn	199
Bill Schindler	22	Granatelli No.1	Grancor Auto Specialists	Kurtis 3000	Offy	13	21	Broken magento strap	132
Jack McGrath	98	Wynn's Friction Proofing No.28	Kurtis Kraft, Inc.	Kurtis 3000	OffySc	52		Rel Agabashn 147-198	0
Jack McGrath	99	Hinkle No.49	Jack B. Hinkle	Kurtis 3000	Offy		0	DNS, too slow	0
Walt Brown	99	Tuffy's Offy No.4	Charles Pritchard	Kurtis 2000	Offy		0	DNS, too slow	0
Kenny Eaton	99	Peter Wales Trucking No.62	M. Peter Wales	Kurtis 2000	Offy		0	DNS, too slow	0
Joe Sostilio	99	Warke No.83	Buster Warke	Kurt\Warke	Offy		0	DNS, too slow	0
Syracuse, NY — 100 Mile — 9 Sep									
Jack McGrath	1	Hinkle No.49	Jack B. Hinkle	Kurtis 3000	Offy	7	100	1:08:41.33 87.350	3,600
Walt Faulkner	2	Grant Piston Ring No.98	J.C. Agajanian	Kurtis 2000	Offy	13	100	Finished	2,550
Cecil Green	4	John Zink No.54	M.A. Walker	Kurtis 3000	Offy	16	100	Finished	975
Fred Agabashian	6	Wynn's Friction Proofing No.28	Kurtis Kraft, Inc.	Kurtis 3000	OffySc	17	100	Finished	750
Ray Knepper	7	Bardahl\Schoof No.86	William Schoof	Kurtis 2000	Offy	15	100	Finished	600
Duke Nalon	8	Chapman No.17	H.A. Chapman	Kurtis 2000	Offy	4	99	Flagged	562
Duke Dinsmore	10	Brown Motor No.69	Verlin Brown	Kurtis 2000	Offy	10	99	Flagged	487
Jerry Hoyt	11	Dr. Morris No.81	Ludson D. Morris, MD	Kurtis 2000	Offy	12	96	Flagged	450
Tony Bettenhausen	12	Belanger No.99	Murrell Belanger	Kurtis	Offy	1	95	Flagged	375
Walt Brown	13	Tuffy's Offy No.4	Charles Pritchard	Kurtis 2000	Offy	3	95	Broken throttle	337
Neal Carter	16	Marion Engr No.23	Milt Marion	Kurtis 2000	OffySc	9	90	Flagged	225
Andy Linden	99	Neal's Restaurant No.62	M. Peter Wales	Kurtis 2000	Offy		0	DNS, too slow	100
Bill Schindler	99	Warke No.83	Buster Warke	Kurt\Warke	Offy		0	DNS, too slow	50
Detroit, MI — 100 Mile — 10 Sep									
Cecil Green	3	John Zink No.54	M.A. Walker	Kurtis 3000	Offy	14	100	Finished	1,836
Walt Faulkner	4	Grant Piston Ring No.98	J.C. Agajanian	Kurtis 2000	Offy	5	100	Finished	1,326
Duke Dinsmore	5	Brown Motor No.69	Verlin Brown	Kurtis 2000	Offy	18	100	Finished	1,224
George Connor	9	Peter Wales Trucking No.62	M. Peter Wales	Kurtis 2000	Offy	16	94	Flagged	714
Tony Bettenhausen	10	Belanger No.99	Murrell Belanger	Kurtis	Offy	2	88	Locked rear end	663
Neal Carter	12	Marion Engr No.23	Milt Marion	Kurtis 2000	OffySc	7	85	Out	510
Duke Nalon	13	Chapman No.17	H.A. Chapman	Kurtis 2000	Offy	4	83	Out of fuel	459
Mack Hellings	14	Tuffy's Offy No.15	Charles Pritchard	Kurtis 2000	Offy	12	74	Out of gas	408
Bill Schindler	15	Granatelli No.11	Grancor Auto Specialists	Kurtis 3000	Offy	8	23	Out	357
Johnnie Parsons	16	Jim Robbins No.1	Jim Robbins	Kurtis	Offy	9	21	Flipped turn 4	306
Jack McGrath	17	Hinkle No.49	Jack B. Hinkle	Kurtis 3000	Offy	13	7	Out	204
Walt Brown	99	Tuffy's Offy No.4	Charles Pritchard	Kurtis 2000	Offy		0	DNS, too slow	0
Ray Knepper	99	Bardahl\Schoof No.86	William Schoof	Kurtis 2000	Offy		0	DNS, too slow	0
Springfield, IL — 100 Mile — 1 Oct									
Tony Bettenhausen	1	Belanger No.99	Murrell Belanger	Kurtis	Offy	1	100	1:08:27.84 87.637	1,800
Fred Agabashian	4	Wynn's Friction Proofing No.28	Kurtis Kraft, Inc.	Kurtis 3000	OffySc	2	100	Finished	488
Cecil Green	5	John Zink No.54	M.A. Walker	Kurtis 3000	Offy	15	100	Finished	450
Ray Knepper	7	Bardahl\Schoof No.86	William Schoof	Kurtis 2000	Offy	16	100	Finished	300
Mack Hellings	9	Tuffy's Offy No.15	Charles Pritchard	Kurtis 2000	Offy	14	98	Flagged	262
Walt Brown	10	Tuffy's Offy No.4	Charles Pritchard	Kurtis 2000	Offy	12	97	Flagged	244
Duke Dinsmore	11	Brown Motor No.69	Verlin Brown	Kurtis 2000	Offy	17	97	Flagged	225
Walt Faulkner	12	Grant Piston Ring No.98	J.C. Agajanian	Kurtis 2000	Offy	10	93	Flagged	188
Jack McGrath	14	Hinkle No.49	Jack B. Hinkle	Kurtis 3000	Offy	8	77	Out	150
Johnnie Parsons	99	Chapman No.17	H.A. Chapman	Kurtis 2000	Offy		0	DNS, too slow	0
Buster Warke	99	Warke No.83	Buster Warke	Kurt\Warke	Offy		0	DNS, too slow	0
Bill Schindler	99	Granatelli No.1	Grancor Auto Specialists	Kurtis 3000	Offy		0	Did not qualify	0
Billy DeVore	99	Federal Engineering No.31	R.A. Cott	Kurtis 3000	Offy		0	Did not qualify	0
Jerry Hoyt	99	Dr. Morris No.81	Ludson D. Morris, MD	Kurtis 2000	Offy		0	Did not qualify	0
Sacramento, CA — 100 Mile — 15 Oct									
Duke Dinsmore	1	Brown Motor No.69	Verlin Brown	Kurtis 2000	Offy	7	100	1:12:45 82.474	2,488
Bill Schindler	2	Chapman No.17	H.A. Chapman	Kurtis 2000	Offy	5	100	Finished	1,763
Fred Agabashian	5	Wynn's Friction Proofing No.28	Kurtis Kraft, Inc.	Kurtis 3000	OffySc	6	100	Finished	622
Mack Hellings	6	Tuffy's Offy No.15	Charles Pritchard	Kurtis 2000	Offy	8	100	Finished	518
Ray Knepper	8	Bardahl\Schoof No.86	William Schoof	Kurtis 2000	Offy	14	99	Flagged	389
Walt Faulkner	9	Grant Piston Ring No.98	J.C. Agajanian	Kurtis 2000	Offy	11	98	Flagged	363

Driver	Fin Pos	Car Name	Entrant	Chassis	Engine	Start Pos	Laps	Time/Speed/Out	Purse
Cecil Green	10	John Zink No.54	M.A. Walker	Kurtis 3000	Offy	12	2	Flagged	332
Johnnie Parsons	11	Esmeralda No.63	Leo Dobry	Kurtis 2000	Offy	17	85	Boken radius rod	311
Jack McGrath	14	Hinkle No.49	Jack B. Hinkle	Kurtis 3000	Offy	13	60	Steering	207
Tony Bettenhausen	17	Belanger No.99	Murrell Belanger	Kurtis	Offy	1	10	Wrecked turn 3	103
Andy Linden	98	John Zink No.54	M.A. Walker	Kurtis 3000	Offy		96	Rel Green 3-96	0
Bill Vukovich	99	Ross Page No.23	Ross Page	Kurtis	OffySc	0		Blew engine practice	0
Phoenix, AZ — 100 Mile — 12 Nov									
Tony Bettenhausen	2	Belanger No.99	Murrell Belanger	Kurtis	Offy	2	100	Finished	1,700
Johnnie Parsons	3	Jim Robbins No.1	Jim Robbins	Kurtis	Offy	6	100	Finished	900
Walt Faulkner	5	Grant Piston Ring No.98	J.C. Agajanian	Kurtis 2000	Offy	7	100	Finished	600
Cecil Green	6	John Zink No.54	M.A. Walker	Kurtis 3000	Offy	9	100	Finished	500
Jack McGrath	9	Hinkle No.49	Jack B. Hinkle	Kurtis 3000	Offy	15	98	Flagged	350
Rodger Ward	10	Esmeralda No.63	Leo Dobry	Kurtis 2000	Offy	16	98	Flagged	325
Mack Hellings	11	Tuffy's Offy No.15	Charles Pritchard	Kurtis 2000	Offy	10	96	Flagged	300
Ray Knepper	13	Bardahl\Schoof No.35	William Schoof	Kurtis 2000	Offy	17	66	Out	225
Duke Dinsmore	18	Brown Motor No.69	Verlin Brown	Kurtis 2000	Offy	11	9	Wrecked turn 1	100
Bill Schindler	99	Chapman No.17	H.A. Chapman	Kurtis 2000	Offy		0	DNS, too slow	0
Johnny Tolan	99	Peter Wates Trucking No.68	M. Peter Wales	Kurtis 2000	Offy		0	DNS, too slow	0
Bay Meadows, CA — 150 Mile — 26 Nov									
Tony Bettenhausen	1	Belanger No.99	Murrell Belanger	Kurtis	Offy	1	149	1:43:45.40 86.163	3,084
Bill Schindler	3	Chapman No.17	H.A. Chapman	Kurtis 2000	Offy	7	149	Finished	1,158
Jack McGrath	5	Hinkle No.49	Jack B. Hinkle	Kurtis 3000	Offy	6	80	Finished	772
Walt Faulkner	6	Grant Piston Ring No.98	J.C. Agajanian	Kurtis 2000	Offy	10	0		643
Duke Dinsmore	7	Brown Motor No.69	Verlin Brown	Kurtis 2000	Offy	13	0		515
Cecil Green	9	John Zink/Walker No.54	M.A. Walker	Kurtis 3000	Offy	5	0		450
Ray Knepper	10	Bardahl\Schoof No.35	William Schoof	Kurtis 2000	Offy	14	0		418
Fred Agabashian	13	Wynn's Friction Proofing No.28	Kurtis Kraft, Inc.	Kurtis 3000	OffySc	15	87	Piston	289
Mack Hellings	98	Hinkle No.49	Jack B. Hinkle	Kurtis 3000	Offy		69	Rel McGrath 81-149	0
Johnny Tolan	99	Peter Wales Trucking No.62	M. Peter Wales	Kurtis 2000	Offy		0	DNS, too slow	0
Mack Hellings	99	Esmeralda No.63	Leo Dobry	Kurtis 2000	Offy		0	DNS, too slow	0
Johnnie Parsons	99	Jim Robbins No.1	Jim Robbins	Kurtis	Offy		0	DNS, too slow	0
Frank Armi	99	Ross Page No.23	Ross Page	Kurtis	OffySc	0		DNS, too slow	0
Darlington, SC — 200 Mile — 10 Dec									
Bill Schindler	2	Chapman No.17	H.A. Chapman	Kurtis 2000	Offy	1	160	1:55:05.38 104.267	3,200
Cecil Green	4	John Zink No.54	M.A. Walker	Kurtis 3000	Offy	3	160	Finished	1,300
Duke Dinsmore	5	Brown Motor No.69	Verlin Brown	Kurtis 2000	Offy	14	160	Finished	1,200
Joie Chitwood	7	Jim Robbins No.1	Jim Robbins	Kurtis	Offy	12	160	Finished	800
Walt Faulkner	8	Grant Piston Ring No.98	J.C. Agajanian	Kurtis 2000	Offy	5	160	Finished	650
Jack McGrath	9	Hinkle No.49	Jack B. Hinkle	Kurtis 3000	Offy	17	160	Finished	600
Jerry Hoyt	11	Dr. Morris No.81	Ludson D. Morris, MD	Kurtis 2000	Offy	20	157	Flagged	500
Ray Knepper	16	Bardahl\Schoof No.35	William Schoof	Kurtis 2000	Offy	26	146	Out	200
Johnny Tolan	17	Peter Wales Trucking No.62	M. Peter Wales	Kurtis 2000	Offy	13	143	Out	200
Walt Brown	19	Federal Engineering No.31	R.A. Cott	Kurtis 3000	Offy	18	138	Spun into pit wall	100
Tony Bettenhausen	20	Belanger No.99	Murrell Belanger	Kurtis	Offy	11	108	Out	100
Duane Carter	21	Grancor No.59	Grancor Auto Specialists	Kurtis 3000	Offy	2	94	Out	100
Mack Hellings	99	Tuffy's Offy No.16	Charles Pritchard	Kurtis 2000	Offy		0	Wrecked practice	0
Williams Grove, PA — 25 Mile — 12 Aug									
Walt Faulkner	4	Grant Piston Ring No.98	J.C. Agajanian	Kurtis 2000	Offy	4	50	Finished	675
Duke Nalon	6	Chapman No.17	H.A. Chapman	Kurtis 2000	Offy	12	50	Finished	450
Neal Carter	7	Marion Engr No.23	Milt Marion	Kurtis 2000	Offy	9	50	Finished	450
Tom Hinnershitz	8	Warke No.83	Buster Warke	Kurt\Warke	Offy	14	50	Finished	337
Mack Hellings	9	Tuffy's Offy No.15	Charles Pritchard	Kurtis 2000	Offy	13	50	Finished	337
George Connor	11	Peter Wales Trucking No.62	M. Peter Wales	Kurtis 2000	Offy	11	50	Finished	337
Jack McGrath	99	Hinkle No.49	Jack B. Hinkle	Kurtis 3000	Offy		0	Did not start	75
Walt Brown	99	Tuffy's Offy No.4	Charles Pritchard	Kurtis 2000	Offy		0	Did not start	112
Duke Dinsmore	99	Brown Motor Co. No.69	Verlin Brown	Kurtis 2000	Offy		0	Did not start	0
Fred Agabashian	99	Wynn's Friction Proofing No.28	Kurtis Kraft, Inc.	Kurtis 3000	Offy		0	Did not start	0
Duane Carter	99	Belanger No.99	Murrell Belanger	Kurtis	Offy		0	Did not start	0
1951									
Indianapolis, IN — 500 Mile — 30 May									
Lee Wallard	1	Belanger No.99	Murrell Belanger	Kurtis	Offy	2	200	3:57:38.05 126.244	63,612
Mike Nazaruk	2	Jim Robbins No.83	Jim Robbins	Kurtis	Offy	7	200	3:59:25.31 125.302	21,362
Jack McGrath	3	Hinkle No.9	Jack Hinkle	Kurtis 3000	Offy	3	100	4:00:29.42 124.745	14,962
Carl Forberg	7	Auto Shippers No.68	Lou Rassey	Kurtis 3000	Offy	24	193	Flagged	5,862
Duke Nalon	10	Novi Purelube No.18	Lou Welch	Kurtis FD	Novi	1	151	Stalled	5,062
Gene Force	11	Brown Motor Co. No.69	Verlin Brown	Kurtis 2000	Offy	22	142	Lost oil pressure	3,182
Sam Hanks	12	Peter Schmidt No.25	Peter Schmidt	Kurtis 3000	Offy	12	135	Spun out NE	3,412
Bill Schindler	13	Chapman No.10	H.A. Chapman	Kurtis 2000	Offy	16	129	Brkn connecting rod	3,142
Fred Agabashian	17	Granatelli Bardahl No.59	Andy Granatelli	Kurtis 3000	Offy	11	109	Clutch	2,862
Carl Scarborough	18	McNamara No.73	Lee Elkins	Kurtis 2000	Offy	15	100	Broken axle	2,642

Driver	Fin Pos	Car Name	Entrant	Chassis	Engine	Start Pos	Laps	Time/Speed/Out	Purse
Johnnie Parsons	21	Wynn's Friction Proofing No.3	Ed Walsh	Kurtis 3000	Offy	8	87	Magneto	2,252
Cecil Green	22	John Zink No.4	M.A. Walker	Kurtis 3000	Offy	10	80	Brkn connecting rod	2,622
Troy Ruttman	23	Agajanian Featherweight No.98	J.C. Agajanian	Kurtis 2000	Offy	6	78	Burned bearing	2,092
Chet Miller	25	Novi Purelube No.32	Lou Welch	Kurtis FD	Novi	28	56	Ignition	2,532
Walt Brown	26	Federal Engr Detroit No.44	Dan LeVine	Kurtis 3000	Offy	13	55	Magneto	2,302
Cliff Griffith	28	Morris No.23	Ludson Morris, MD	Kurtis 2000	Offy	18	30	Broken rear axle	2,042
Manuel Ayulo	98	Hinkle No.9	Jack Hinkle	Kurtis 3000	Offy		100	Rel. McGrath 101-200	0
Bob Sweikert	99	Marion Engineering No.37	Milt Marion	Kurtis 2000	Offy		0	DNS, too slow	0
Frank Armi	99	Bardahl\Page No.64	Dick Page	Kurtis	Duray	0		DNS, too slow	0
Johnny Tolan	99	Pete Wales Trucking No.34	Peter Wales	Kurtis 2000	Offy		0	Did not quality	0
LeRoy Warriner	99	Heller No.75	George F. Heller	Kurtis	Offy		0	Did not qualify	0
Bill Cantrell	99	Hart Fullerton No.79	Hart Fullerton	Kurtis 4000	OffySc	0		Did not qualify	0

Milwaukee, WI — 100 Mile — 10 Jun

Driver	Fin Pos	Car Name	Entrant	Chassis	Engine	Start Pos	Laps	Time/Speed/Out	Purse
Tony Bettenhausen	1	Belanger No.99	Murrell Belanger	Kurtis	Offy	5	100	1:06:38.91 90.024	4,500
Cecil Green	2	John Zink No.4	M.A. Walker	Kurtis	Offy	6	100	Finished	3,240
Gene Force	7	Brown Motor Co. No.69	Verlin Brown	Kurtis 2000	Offy	13	98	Flagged	720
Mike Nazaruk	8	Jim Robbins No.83	Jim Robbins	Kurtis	Offy	18	98	Flagged	720
Cliff Griffith	9	Morris No.23	Ludson Morris, MD	Kurtis 2000	Offy	7	98	Flagged	540
Johnny Tolan	12	Pete Wales Trucking No.34	Peter Wales	Kurtis 2000	Offy	15	94	Flagged	450
Bill Schindler	16	Chapman No.10	H.A. Chapman	Kurtis 2000	Offy	12	20	Magneto trouble	360
Jack McGrath	18	Hinkle No.9	Jack Hinkle	Kurtis 4000	Offy	1	0	Spun out turn 1	180
George Connor	99	Agajanian Featherweight No.98	J.C. Agajanian	Kurtis 2000	Offy		0	DNS, too slow	0
Joe Sostilio	99	Ess-A-R-R Engineering No.79	S.T. Richter	Kurt Midgt	Offy		0	DNS, too slow	0
Johnnie Parsons	99	Wynn's Friction Proofing No.3	Ed Walsh	Kurtis 3000	Offy		0	DNS, too slow	0
Fred Agabashian	99	Granatelli No.59	Andy Granatelli	Kurtis 3000	Offy		0	DNS, too slow	0
Frank Armi	99	Bardahl\Page No.64	Dick Page	Kurtis	Offy		0	DNS, too slow	0
Carl Scarborough	99	McNamara No.73	Lee Elkins	Kurtis 2000	Offy		0	Did not qualify	0
Jud Larson	99	Cocker No.75		Kurtis	Offy		0	Did not qualify	0

Langhorne, PA — 100 Mile — 24 Jun

Driver	Fin Pos	Car Name	Entrant	Chassis	Engine	Start Pos	Laps	Time/Speed/Out	Purse
Tony Bettenhausen	1	Belanger No.99	Murrell Belanger	Kurtis	Offy	5	100	1:04:57.18 92.357	2,375
Bill Schindler	2	Chapman No.10	H.A. Chapman	Kurtis 2000	Offy	3	100	Finished	1,696
Cecil Green	3	John Zink No.4	M.A. Walker	Kurtis	Offy	11	100	Finished	1,014
Fred Agabashian	6	Granatelli No.59	Andy Granatelli	Kurtis 3000	Offy	15	98	Flagged	422
Jack McGrath	7	Hinkle No.9	Jack Hinkle	Kurtis 4000	Offy	1	98	Flagged	341
Mike Nazaruk	9	Jim Robbins No.83	Jim Robbins	Kurtis	Offy	17	71	Flagged	341
Cliff Griffith	11	Morris No.23	Ludson Morris, MD	Kurtis 2000	Offy	2	84	Flagged	257
Johnny McDowell	16	Hart Fullerton No.79	Hart Fullerton	Kurtis 4000	OffySc	6	8	Overheated	171
Johnnie Parsons	98	Jim Robbins No.83	Jim Robbins	Kurtis	Offy		26	Rel Nazaruk72-97	0
Jud Larson	99	Cocker No.75		Kurtis	Offy		0	DNS, too slow	0
Tom Hinnershitz	99	Warke No.88	Buster Warke	Kurt\Warke	Offy		0	DNS, too slow	0
Joe Sostilio	99	Ess-A-R-R Engineering No.79	S.T. Richter	Kurt Midgt	Offy		0	DNS, too slow	0
Mack Hellings	99	Derrico No.19	J. Derrico	Kurtis 2000	Offy		0	Did not qualify	0
George Connor	99	Marion Engineering No.37	Milt Marion	Kurtis 2000	Offy		0	Broken rear axle	0
Carl Scarborough	99	McNamara No.73	Lee Elkins	Kurtis 2000	Offy		0	Magneto	0_

Darlington, SC — 250 Mile — 4 Jul

Driver	Fin Pos	Car Name	Entrant	Chassis	Engine	Start Pos	Laps	Time/Speed/Out	Purse
Tony Bettenhausen	2	Belanger No.99	Murrell Belanger	Kurtis	Offy	12	200	2:24:26.5 103.848	3,594
Cecil Green	3	John Zink No.4	M.A. Walker	Kurtis	Offy	2	200	Finished	2,246
Jack McGrath	5	Hinkle No.9	Jack Hinkle	Kurtis 3000	Offy	14	200	Finished	1,347
Mike Nazaruk	8	Jim Robbins No.83	Jim Robbins	Kurtis	Offy	18	199	Flagged	730
Walt Brown	10	Federal Engr Detroit No.44	Dan LeVine	Kurtis 3000	Offy	3	195	Flagged	673
Cliff Griffith	11	Morris No.23	Ludson Morris, MD	Kurtis 2000	Offy	10	152	Flagged	561
Kenny Eaton	14	Hinkle No.19	Jack Hinkle	Kurtis 4000	Offy	9	181	Flagged	449
Gene Force	15	Brown Motor Co. No.69	Verlin Brown	Kurtis 2000	Offy	26	107	Flagged	224
Johnny McDowell	16	Hart Fullerton No.79	Hart Fullerton	Kurtis 4000	OffySc	1	170	Flagged	224
Bill Schindler	21	Chapman No.10	H.A. Chapman	Kurtis 2000	Offy	16	117	Magneto	112
Johnnie Parsons	23	Marion Engineering No.37	Milt Marion	Kurtis 2000	Offy	20	85	Brkn rear end	112
George Connor	25	Sarafoff No.98	Tom Sarafoff	Kurtis 2000	Offy	8	45	Brkn piston	112
Carl Scarborough	27	McNamara No.73	Lee Elkins	Kurtis 2000	Offy	11	17	Brkn oil line	112
Rodger Ward	98	Morris No.23	Ludson Morris, MD	Kurtis 2000	Offy		41	Rel. Griffith 153-93	0
Bill Vukovich	98	Federal Engr Detroit No.44	Dan LeVine	Kurtis 3000	Offy		0	Rel. Brown	0
Joe James	98	Brown Motor Co. No.69	Verlin Brown	Kurtis 2000	Offy		70	Rel. Force 83-152	0
Carl Forberg	99	Auto Shippers No.68	Lou Rassey	Kurtis 3000	Offy		0	DNS, too slow	0
Fred Agabashian	99	Granatelli Bardahl No.59	Andy Granatelli	Kurtis 3000	Offy		0	DNS, too slow	0
Johnnie Parsons	99	Wynn's Friction Proofing No.3	Ed Walsh	Kurtis 3000	Offy		0	DNS, too slow	0
Johnny Tolan	99	Pete Wales Trucking No.34	Peter Wales	Kurtis 2000	Offy		0	Did not quality	0

Springfield, IL — 100 Mile — 18 Aug

Driver	Fin Pos	Car Name	Entrant	Chassis	Engine	Start Pos	Laps	Time/Speed/Out	Purse
Tony Bettenhausen	1	Belanger No.99	Murrell Belanger	Kurtis	Offy	14	100	1:06:37.59 90.054	2,879
Jack McGrath	2	Hinkle No.9	Jack Hinkle	Kurtis 4000	Offy	4	100	Finished	2,070
Bill Schindler	3	Chapman No.10	H.A. Chapman	Kurtis 2000	Offy	7	100	Finished	1,150
Troy Ruttman	6	John Zink No.4	M.A. Walker	Kurtis	Offy	12	100	Finished	575
Gordon Reid	7	McNamara No.73	Lee Elkins	Kurtis 2000	Offy	17	98	Flagged	460

Driver	Fin Pos	Car Name	Entrant	Chassis	Engine	Start Pos	Laps	Time/Speed/Out	Purse
Johnny McDowell	9	Hart Fullerton No.79	Hart Fullerton	Kurtis 4000	OffySc	15	97	Flagged	402
Gene Force	13	Brown Motor Co. No.69	Verlin Brown	Kurtis 2000	Offy	11	81	Steering failure	258
Harry King	99	Pete Wales Trucking No.34	Peter Wales	Kurtis 2000	Offy		0	DNS, too slow	0
George Connor	99	Saraffof No.98	Tom Sarafoff	Kurtis 2000	Offy		0	Did not qualify	0
Milwaukee, WI — 200 Mile — 26 Aug									
Jack McGrath	2	Hinkle No.9	Jack Hinkle	Kurtis 4000	Offy	6	200	Finished	4,247
Johnnie Parsons	3	Wynn's Friction Proofing No.3	Ed Walsh	Kurtis 3000	Offy	12	200	Finished	2,831
Bill Schindler	4	Chapman No.10	H.A. Chapman	Kurtis 2000	Offy	9	200	Finished	2,123
Cliff Griffith	7	Sarafoff No.98	Tom Sarafoff	Kurtis 2000	Offy	15	196	Flagged	1,132
Johnny McDowell	8	Hart Fullerton No.79	Hart Fullerton	Kurtis 4000	OffySc	1	195	Flagged	1,062
Tony Bettenhausen	10	Belanger No.99	Murrell Belanger	Kurtis	Offy	4	187	Flagged	920
Duke Dinsmore	18	Federal Engr Detroit No.44	Dan LeVine	Kurtis 3000	Offy	22	81	Accident turn 1	283
Gene Force	19	Brown Motor Co. No.69	Verlin Brown	Kurtis 2000	Offy	13	43	Engine failure	283
Gordon Reid	20	McNamara No.73	Lee Elkins	Kurtis 2000	Offy	16	33	Magneto	283
Bill Vukovich	21	John Zink No.4	M.A. Walker	Kurtis	Offy	3	30	Brkn connecting rod	142
Harry King	99	Pete Wales Trucking No.34	Peter Wales	Kurtis 2000	Offy		0	DNS, too slow	0
DuQuoin, IL — 100 Mile — 1 Sep									
Tony Bettenhausen	1	Belanger No.99	Murrell Belanger	Kurtis	Offy	6	100	1:07:56.48 88.311	1,875
Johnnie Parsons	2	Wynn's Friction Proofing No.3	Ed Walsh	Kurtis 3000	Offy	1	100	Finished	1,350
Bill Vukovich	7	John Zink No.4	M.A. Walker	Kurtis	Offy	15	98	Flagged	300
Cliff Griffith	8	Sarafoff No.98	Tom Sarafoff	Kurtis 2000	Offy	13	98	Flagged	281
Bill Schindler	12	Chapman No.10	H.A. Chapman	Kurtis 2000	Offy	10	94	Flagged	187
Johnny McDowell	15	Hart Fullerton No.79	Hart Fullerton	Kurtis 4000	OffySc	14	85	Supercharger trouble	131
Jack McGrath	16	Hinkle No.9	Jack Hinkle	Kurtis 4000	Offy	3	71	Bent valve	112
Gordon Reid	18	McNamara No.73	Lee Elkins	Kurtis 2000	Offy	8	40	Broken gear	75
Harry King	99	Pete Wales Trucking No.34	Peter Wales	Kurtis 2000	Offy		0	DNS, too slow	0
DuQuoin, IL — 200 Mile — 3 Sep									
Tony Bettenhausen	1	Belanger No.99	Murrell Belanger	Kurtis	Offy	1	101	1:09:19.00 86.559	3,100
Harry King	8	Pete Wales Trucking No.34	Peter Wales	Kurtis 2000	Offy	16	97	Flagged - rain	465
Cliff Griffith	9	Sarafoff No.98	Tom Sarafoff	Kurtis 2000	Offy	6	96	Flagged - rain	434
Gene Force	10	Brown Motor Co. No.69	Verlin Brown	Kurtis 2000	Offy	11	93	Flagged - rain	403
Jack McGrath	11	Hinkle No.9	Jack Hinkle	Kurtis 4000	Offy	18	93	Flagged - rain	372
Gordon Reid	13	McNamara No.73	Lee Elkins	Kurtis 2000	Offy	3	87	Brkn radius rod	279
Johnny McDowell	15	Hart Fullerton No.79	Hart Fullerton	Kurtis 4000	OffySc	12	71	Stalled	217
Johnnie Parsons	16	Wynn's Friction Proofing No.3	Ed Walsh	Kurtis 4000	Offy	15	37	Wrecked, flip turn 2	186
Bill Vukovich	99	John Zink No.4	M.A. Walker	Kurtis	Offy		0	DNS, too slow	0
Pikes Peak, CO — 12 Mile — 4 Sep									
Johnny Mauro	2	Dobry No.63	Leo Dobry	Kurtis 2000	Offy	3	1	0:15:47.05 47.212	2,500
Louis Unser	3	Federal Engr Detroit No.44	Dan LeVine	Kurtis 3000	Offy	1	1	0:16:06.85 46.245	1,500
Syracuse, NY — 100 Mile — 8 Sep									
Tony Bettenhausen	1	Belanger No.99	Murrell Belanger	Kurtis	Offy	1	67	Flagged, wreck	3,750
Bill Vukovich	3	John Zink No.4	M.A. Walker	Kurtis	Offy	6	67	Flagged, wreck	1,500
Harry King	5	Pete Wales Trucking No.34	Peter Wales	Kurtis 2000	Offy	18	66	Flagged, wreck	900
Jack McGrath	7	Hinkle No.9	Jack Hinkle	Kurtis 4000	Offy	7	65	Wrecked, turn 2	600
Bill Schindler	8	Chapman No.10	H.A. Chapman	Kurtis 2000	Offy	11	65	Wrecked, turn 2	562
Buzz Barton	10	Ess-A-R-R Engineering No.79	S.T. Richter	Kurt Midgt	Offy	10	64	Wrecked, turn 2	417
Tom Hinnershitz	14	Warke No.83	Buster Warke	Kurt\Warke	Offy	12	33	Lost brakes	300
Johnnie Parsons	17	Marion Engineering No.37	Milt Marion	Kurtis 2000	Offy	13	6	Brkn connecting rod	150
Detroit, MI — 100 Mile — 9 Sep									
Tony Bettenhausen	4	Belanger No.99	Murrell Belanger	Kurtis	Offy	4	100	Finished	845
Bill Vukovich	7	John Zink No.4	M.A. Walker	Kurtis	Offy	13	100	Finished	520
Buster Warke	12	Warke No.83	Buster Warke	Kurt\Warke	Offy	17	85	Flagged	325
Gordon Reid	14	McNamara No.73	Lee Elkins	Kurtis 2000	Offy	8	33	Stripped gear	260
Johnny McDowell	17	Hart Fullerton No.79	Hart Fullerton	Kurtis 4000	OffySc	1	20	Blown engine	150
Gene Force	18	Brown Motor Co. No.69	Verlin Brown	Kurtis 2000	Offy	5	6	Brkn connecting rod	150
Duke Dinsmore	99	Federal Engr Detroit No.44	Dan LeVine	Kurtis 3000	Offy		0	DNS, too slow	0
Harry King	99	Pete Wales Trucking No.34	Peter Wales	Kurtis 2000	Offy		0	DNS, too slow	0
Denver, CO — 100 Mile — 23 Sep									
Tony Bettenhausen	1	Belanger No.99	Murrell Belanger	Kurtis	Offy	1	100	1:09:15.00 86.640	3,537
Jack McGrath	3	Hinkle No.9	Jack Hinkle	Kurtis 4000	Offy	12	100	Finished	1,415
Joe James	4	John Zink No.4	M.A. Walker	Kurtis	Offy	11	100	Finished	919
Bobby Ball	5	Blakely Oil No.52	John L. McDaniel	Kurtis 4000	Offy	15	100	Finished	849
Johnny McDowell	9	Federal Engr Detroit No.44	Dan LeVine	Kurtis 3000	Offy	8	95	Flagged	495
Johnnie Parsons	14	Wynn's Friction Proofing No.3	Ed Walsh	Kurtis 4000	Offy	10	77	Lost brakes	283
Gordon Reid	16	McNamara No.73	Lee Elkins	Kurtis 2000	Offy	9	65	Brkn shock absorber	212
Del Fanning	99	Dobry No.63	Leo Dobry	Kurtis 2000	Offy		0	DNS, too slow	0

Driver	Fin Pos	Car Name	Entrant	Chassis	Engine	Start Pos	Laps	Time/Speed/Out	Purse
San Jose, CA — 100 Mile — 21 Oct									
Tony Bettenhausen	1	Belanger No.99	Murrell Belanger	Kurtis	Offy	1	100	1:14:24.00 81.760	1,979
Jack McGrath	4	Hinkle No.9	Jack Hinkle	Kurtis 4000	Offy	6	98	Flagged	514
Johnnie Parsons	7	Wynn's Friction Proofing No.3	Ed Walsh	Kurtis 4000	Offy	5	96	Flagged	316
Joe James	8	John Zink No.4	M.A. Walker	Kurtis	Offy	17	91	Flagged	296
Johnny McDowell	9	Federal Engr Detroit No.44	Dan LeVine	Kurtis 3000	Offy	18	92	Flagged	277
Bobby Ball	11	Blakely Oil No.52	John L. McDaniel	Kurtis 4000	Offy	16	84	Flagged	237
Gordon Reid	13	McNamara No.73	Lee Elkins	Kurtis 2000	Offy	11	70	Brkn radius rod	178
Neal Carter	14	Lutes No.28	W.J. Lutes	Kurtis 4000	Offy	7	50	Brkn engine timing	158
Troy Ruttman	15	Hart Fullerton No.79	Hart Fullerton	Kurtis 4000	OffySc	15	39	Brkn steering	138
Gordon Reid	98	John Zink No.4	M.A. Walker	Kurtis	Offy		4	Rel. James 92-95	0
George Connor	99	Chapman No.10	H.A. Chapman	Kurtis 2000	Offy		0	DNS, too slow	0
Del Fanning	99	Dobry No.63	Leo Dobry	Kurtis 2000	Offy		0	DNS, too slow	0
Gene Force	99	Brown Motor Co. No.69	Verlin Brown	Kurtis 2000	Offy		0	DNS, too slow	0
Dick Page	99	Bardahl\Page No.64	Dick Page	Kurtis	Offy		0	DNS, too slow	0
Dempsey Wilson	99	Johnston No.55	Fred Johnston	Kurt Midgt	Offy		0	Did not qualify	0
Phoenix, AZ — 100 Mile — 4 Nov									
Johnnie Parsons	1	Wynn's Friction Proofing No.3	Ed Walsh	Kurtis 4000	Offy	8	100	1:10:54.00 84.626	2,584
Neal Carter	4	Lutes No.28	W.J. Lutes	Kurtis 4000	Offy	3	99	Flagged	671
Joe James	5	John Zink No.4	M.A. Walker	Kurtis	Offy	12	99	Flagged	620
Gordon Reid	7	McNamara No.73	Lee Elkins	Kurtis 2000	Offy	10	98	Flagged	413
Bill Schindler	8	Chapman No.10	H.A. Chapman	Kurtis 2000	Offy	16	97	Flagged	387
Johnny McDowell	12	Federal Engr Detroit No.44	Dan LeVine	Kurtis 3000	Offy	18	90	Flagged	258
Tony Bettenhausen	13	Belanger No.99	Murrell Belanger	Kurtis	Offy	4	90	Blew tire	232
Jack McGrath	14	Hinkle No.9	Jack Hinkle	Kurtis 4000	Offy	7	89	Wrecked north turn	206
Bobby Ball	15	Hart Fullerton No.79	Hart Fullerton	Kurtis 4000	OffySc	5	77	Brkn throttle shaft	180
Gene Force	99	Brown Motor Co. No.69	Verlin Brown	Kurtis 2000	Offy		0	DNS, too slow	0
Mack Hellings	99	Blakely Oil No.52	John L. McDaniel	Kurtis 4000	Offy		0	DNS, too slow	0
Bay Meadows, CA — 150 Mile — 11 Nov									
Johnnie Parsons	1	Wynn's Friction Proofing No.3	Ed Walsh	Kurtis 4000	Offy	13	150	1:42:36.80 87.700	2,500
Tony Bettenhausen	2	Belanger No.99	Murrell Belanger	Kurtis	Offy	2	150	Finished	1,800
Neal Carter	4	Lutes No.28	W.J. Lutes	Kurtis 4000	Offy	4	149	Flagged	650
Gene Force	6	Brown Motor Co. No.69	Verlin Brown	Kurtis 2000	Offy	12	147	Flagged	500
Bill Schindler	9	Chapman No.10	H.A. Chapman	Kurtis 2000	Offy	10	146	Flagged	350
Bobby Ball	13	Hart Fullerton No.79	Hart Fullerton	Kurtis 4000	OffySc	14	137	Flagged	225
Gordon Reid	14	McNamara No.73	Lee Elkins	Kurtis 2000	Offy	16	137	Flagged	200
Jack McGrath	18	Hinkle No.9	Jack Hinkle	Kurtis 4000	Offy	1	23	Burned piston	100
Bill Vukovich	99	Dobry No.63	Leo Dobry	Kurtis 2000	Offy		0	DNS, too slow	0
Joe James	99	John Zink No.4	M.A. Walker	Kurtis	Offy		0	DNS, too slow	0
Williams Grove, PA — 25 Mile — 29 Jul									
Tony Bettenhausen	2	Belanger No.99	Murrell Belanger	Kurtis	Offy	4	50	Finished	975
Tom Hinnershitz	4	Warke No.83	Buster Warke	Kurt\Warke	Offy	7	50	Finished	675
Jack McGrath	5	Hinkle No.9	Jack Hinkle	Kurtis 4000	Offy	12	50	Finished	525
Cliff Griffith	13	Morris No.23	Ludson Morris, MD	Kurtis 2000	Offy	6	50	Finished	225
Bill Schindler	14	Chapman No.10	H.A. Chapman	Kurtis 2000	Offy	14	50	Finished	150
Johnny Tolan	99	Pete Wales Trucking No.34	Peter Wales	Kurtis 2000	Offy		0	Did not start	112
1952									
Indianapolis, IN — 500 Mile — 30 May									
Jim Rathmann	2	Grancor-Wynn No.59	Grancor Auto Specialists	Kurtis 3000	Offy	10	200	3:56:44.21 126.723	24,368
Sam Hanks	3	Bardahl No.18	Ed Walsh	Kurtis 3000	Offy	5	200	3:58:53.49 125.803	14,768
Art Cross	5	Bowes Seal Fast No.33	Ray T. Brady	Kurtis 4000	Offy	20	200	4:01:12.03 124.292	9,218
Jimmy Bryan	6	Peter Schmidt No.77	Peter Schmidt	Kurtis 3000	Offy	21	200	4:02:06.20 123.914	7,468
Jimmy Reece	7	John Zink No.37	John Zink	Kurtis 4000	Offy	23	200	4:03:17.12 123.312	6,368
George Connor	8	Federal Engineering No.54	Federal Automotive Associates	Kurtis 3000	Offy	14	200	4:04:42.49 122.595	6,118
Cliff Griffith	9	Tom Sarafoff No.22	Tom Sarafoff	Kurtis 2000	Offy	9	200	4:05:04.44 122.412	5,768
Johnnie Parsons	10	Jim Robbins No.5	J. M. Robbins	Kurtis	Offy	31	200	4:06:19.66 121.789	5,518
Jack McGrath	11	Hinkle No.4	Jack B. Hinkle	Kurtis 3000	Offy	3	200	4:07:03.60 121.428	4,263
Joe James	13	Bardahl No.14	Ed Walsh	Kurtis 4000	Offy	16	200	4:09:37.53 120.180	2,923
Bill Vukovich	17	Fuel Injection No.26	Howard Keck Co.	Kurtis 500A	Offy	8	191	Wrecked NE	18,693
Chuck Stevenson	18	Springfield Welding No.16	Bessie Lee Paoli	Kurtis 4000	Offy	11	187	Flagged	2,623
Johnny McDowell	21	McDowell No.31	Roger G. Wolcott	Kurtis	Offy	33	182	Flagged	2,333
Rodger Ward	23	Federal Engineering No.34	Federal Automotive Associates	Kurtis 4000	Offy	22	130	Lost oil pressure	2,273
Duke Nalon	25	Novi Pure Oil No.36	Lewis W. Welch	Kurtis FD	NoviSc	4	84	Brkn superchgr shaft	
Bob Sweikert	26	McNamara No.73	Lee Elkins	Kurtis 2000	Offy	32	77	Differential gear	2,183
Fred Agabashian	27	Cummins Diesel No.28	Cummins Engine Co.	Kurtis R	CummTC	1	71	Clogged turbocharger	2,653
Gene Hartley	28	Mel-Rae No.67	Mel B. Wiggers	Kurtis 4000	Offy	18	65	Broken exhaust pipe	2,123
Bob Scott	29	Dr. Morris No.93	Ludson D. Morris, MD	Kurtis 2000	Offy	25	49	Broken drive shaft	2,093
Chet Miller	30	Novi Pure Oil No.21	Lewis W. Welch	Kurtis FD	NoviSc	27	41	Brkn superchgr shaft	3,663
Andy Linden	33	Miracle Power No.9	Hart Fullerton	Kurtis 4000	OffySc	2	20	Oil leak	2,273

Driver	Fin Pos	Car Name	Entrant	Chassis	Engine	Start Pos	Laps	Time/Speed/Out	Purse
Mike Nazaruk	99	John Zink No.66	M. A. Walker	Kurtis 3000	Offy		0	DNS, too slow	0
Jimmy Jackson	99	Auto Shippers No.61	Eugene A. Casaroll	Kurtis 3000	Offy		0	DNS, too slow	0
George Tichenor	99	Schmidt No.88	Peter Schmidt	Kurtis 4000	Offy		0	DNS, too slow	0
Paul Russo	99	Lutes Truck Parts No.10	W.J. Lutes	Kurtis 4000	Offy		0	Did not qualify	0
Neal Carter	99	Auto Shippers No.25	Eugene A. Casaroll	Kurtis 500A	Offy		0	Did not qualify	0
Allen Heath	99	Engle-Stanko No.32	John T. Stanko	Kurtis 4000	Offy		0	Wrecked qual, NW	0
Jimmy Daywalt	99	Merkler Machine Works No.64	Franklin Merkler	Kurtis	OffySc		0	Did not qualify	0
Bayliss Levrett	99	Brown Motor No.69	Brown Motor Co.	Kurtis 2000	Offy		0	Wrecked practice MS	0
Bob Sweikert	99	Marion Engineering No.71	Milt Marion	Kurtis 2000	Offy		0	Did not qualify	0
Johnny Fedricks	99	Cal Connell No.82	Calvin C. Connell	Kurtis	Cadilc		0	Did not arrive	0
Tony Bettenhausen	99	Belanger No.99	Murrell Belanger	Kurtis	Offy		0	Wrecked qual, SW	0
Milwaukee, WI — 100 Mile — 8 Jun									
Mike Nazaruk	1	McNamara No.73	Lee Elkins	Kurtis 2000	Offy	8	100	1:05:02.22 92.255	4,563
Paul Russo	2	Lutes Truck Parts No.10	W.J. Lutes	Kurtis 4000	Offy	15	100	Finished	2,918
Jimmy Reece	5	John Zink No.37	John Zink	Kurtis 4000	Offy	12	98	Flagged	1,095
Chuck Stevenson	6	Springfield Welding No.16	Bessie Lee Paoli	Kurtis 4000	Offy	5	98	Flagged	913
Gene Hartley	7	Mel-Rae No.67	Mel B. Wiggers	Kurtis 4000	Offy	3	97	Flagged	730
Bob Scott	9	Dr. Morris No.93	Ludson D. Morris, MD	Kurtis 2000	Offy	11	96	Flagged	548
Bobby Ball	11	Blakely Oil No.15	John McDaniel	Kurtis 4000	Offy	9	96	Flagged	456
Jack McGrath	13	Hinkle No.4	Jack B. Hinkle	Kurtis 4000	Offy	2	83	Broken oil line	365
Andy Linden	14	Miracle Power No.9	Hart Fullerton	Kurtis 4000	OffySc	4	62	Engine trouble	365
Cliff Griffith	15	Tom Sarafoff No.22	Tom Sarafoff	Kurtis 2000	Offy	7	56	Spun out	365
Allen Heath	16	Engle-Stanko No.32	John T. Stanko	Kurtis 4000	Offy	13	56	Engine trouble	365
Jud Larson	99	John Zink No.66	M. A. Walker	Kurtis	Offy		0	DNS, too slow	25
Tony Bettenhausen	99	Belanger No.99	Murrell Belanger	Kurtis	Offy		0	DNS, too slow	25
Johnnie Parsons	99	Schmidt No.6		Kurtis 4000	Offy		0	DNS, too slow	25
Sam Hanks	99	Bardahl No.14	Ed Walsh	Kurtis 4000	Offy		0	DNS, too slow	25
Joe James	99	Bowes Seal Fast No.33	Ray T. Brady	Kurtis 4000	Offy		0	DNS, too slow	25
Neal Carter	99	Auto Shippers No.25	Eugene A. Casaroll	Kurtis 500A	Offy		0	DNS, too slow	0
Rodger Ward	99	Federal Engineering No.34	Federal Automotive Assoc.	Kurtis 4000	Offy		0	DNS, too slow	0
Johnny McDowell	99	McDowell No.31	Roger G. Wolcott	Kurtis	Offy		0	Wrecked pract-fatal	0
Raleigh, NC — 200 Mile — 4 Jul									
Jack McGrath	2	Hinkle No.4	Jack B. Hinkle	Kurtis 4000	Offy	8	200	Finished	2,800
Jim Rathmann	4	Belanger No.99	Murrell Belanger	Kurtis	Offy	12	200	Finished	1,070
Paul Russo	5	Lutes Truck Parts No.10	W.J. Lutes	Kurtis 4000	Offy	15	199	Flagged	988
Chuck Stevenson	6	Springfield Welding No.16	Bessie Lee Paoli	Kurtis 4000	Offy	16	198	Flagged	864
Neal Carter	7	Auto Shippers No.25	Eugene A. Casaroll	Kurtis 500A	Offy	20	198	Flagged	658
Rodger Ward	9	Federal Engineering No.34	Federal Automotive Assoc.	Kurtis 4000	Offy	14	195	Out of fuel	494
Cliff Griffith	11	Tom Sarafoff No.22	Tom Sarafoff	Kurtis 2000	Offy	1	190	Ignition	411
Danny Oakes	12	John Zink No.66	M. A. Walker	Kurtis	Offy	19	186	Magneto	411
Bob Scott	13	Dr. Morris No.93	Ludson D. Morris, MD	Kurtis 2000	Offy	16	185	Out of fuel	329
Andy Linden	16	Miracle Power No.9	Hart Fullerton	Kurtis 4000	OffySc	10	97	Carburetor	164
Sam Hanks	18	Bardahl No.14	Ed Walsh	Kurtis 4000	Offy	7	89	Broken fuel pump	164
Mike Nazaruk	19	McNamara No.73	Lee Elkins	Kurtis 2000	Offy	18	78	Clutch	82
George Connor	21	Federal Engineering No.54	Joseph J. Barzda	Kurtis 3000	Offy	23	73	Magneto	82
Johnnie Parsons	22	Ricketts No.6		Kurtis 4000	Offy	5	52	Throttle	82
Bill Vukovich	24	Fuel Injection No.26	Howard Keck Co/	Kurtis 500A	Offy	11	36	Lost brakes	82
Bobby Ball	99	Blakely Oil No.15	John McDaniel	Kurtis 4000	Offy		0	Did not qualify	0
Allen Heath	99	Engle-Stanko No.32	John T. Stanko	Kurtis 4000	Offy		0	Did not qualify	0
Joe James	99	Bowes Seal Fast No.33	Ray T. Brady	Kurtis 4000	Offy		0	Did not qualify	0
Jimmy Reece	99	John Zink No.37	John Zink	Kurtis 4000	Offy		0	Did not qualify	0
Gene Hartley	99	Mel-Rae No.67	Mel B. Wiggers	Kurtis 4000	Offy		0	Did not qualify	0
Springfield, IL — 100 Mile — 16 Aug									
Jack McGrath	2	Hinkle No.4	Jack B. Hinkle	Kurtis 4000	Offy	2	100	Finished	2,312
Sam Hanks	3	Bardahl No.14	Ed Walsh	Kurtis 4000	Offy	18	100	Finished	1,589
Chuck Stevenson	5	Springfield Welding No.16	Bessie Lee Paoli	Kurtis 4000	Offy	17	100	Finished	867
Joe James	6	Bowes Seal Fast No.33	Ray T. Brady	Kurtis 4000	Offy	13	100	Finished	722
Jimmy Reece	8	John Zink No.44	John Ziink	Kurtis 4000	Offy	10	99	Flagged	578
Neal Carter	10	Auto Shippers No.25	Eugene A. Casaroll	Kurtis 500A	Offy	8	97	Flagged	433
Gene Hartley	14	Mel-Rae No.67	Mel B. Wiggers	Kurtis 4000	Offy	6	82	Flagged	289
Bobby Ball	15	Blakely Oil No.15	John McDaniel	Kurtis 4000	Offy	12	56	Spun out, hit wall	289
Mike Nazaruk	18	McNamara No.73	Lee Elkins	Kurtis 2000	Offy	1	25	Broken drive shaft	144
Johnnie Parsons	99	Ricketts No.6		Kurtis 4000	Offy		0	DNS, too slow	0
Frank Luptow	99	Dillon No.19		Kurtis 4000	Offy		0	DNS, too slow	0
Andy Linden	99	Miracle Power No.9	Hart Fullerton	Kurtis 4000	OffySc		0	DNS, too slow	0
Jim Rathmann	99	Belanger No.99	Murrell Belanger	Kurtis	Offy		0	DNS, too slow	0
Bill Vukovich	99	Fuel Injection No.26	Howard Keck Co.	Kurtis 500A	Offy		0	DNS, too slow	0
Paul Russo	99	Lutes Truck Parts No.10	W.J. Lutes	Kurtis 4000	Offy		0	DNS, too slow	0
Eddie Russo	99	Pritchard No.19	Charles Pritchard	Kurtis 2000	Offy		0	DNS, too slow	0
Bob Sweikert	99	Dr. Morris No.93	Ludson D. Morris, MD	Kurtis 2000	Offy		0	DNS, too slow	0
Duke Dinsmore	99	Rouge Motor No.32	John T. Stanko	Kurtis 4000	Offy		0	DNS, too slow	0
Allen Heath	99	John Zink No.66	M. A. Walker	Kurtis	Offy		0	DNS, too slow	0

Driver	Fin Pos	Car Name	Entrant	Chassis	Engine	Start Pos	Laps	Time/Speed/Out	Purse
Rodger Ward	99	Federal Engineering No.34	Federal Automotive Assoc.	Kurtis 4000	Offy		0	Did not qualify	0
Milwaukee, WI — 200 Mile — 24 Aug									
Chuck Stevenson	1	Springfield Welding No.16	Bessie Lee Paoli	Kurtis 4000	Offy	11	200	2:27:26.115 81.392	7,440
Joe James	5	Bowes Seal Fast No.33	Ray T. Brady	Kurtis 4000	Offy	13	200	Finished	1,786
Paul Russo	6	Lutes Truck Parts No.10	W.J. Lutes	Kurtis 4000	Offy	16	199	Flagged	1,117
Rodger Ward	7	Federal Engineering No.34	Federal Automotive Assoc.	Kurtis 4000	Offy	20	199	Flagged	1,488
Jimmy Reece	8	John Zink No.44	John Ziink	Kurtis 4000	Offy	15	198	Flagged	1,190
Bobby Ball	10	Blakely Oil No.15	John McDaniel	Kurtis 4000	Offy	2	195	Flagged	967
Sam Hanks	14	Bardahl No.14	Ed Walsh	Kurtis 4000	Offy	3	181	Flagged	595
Jack McGrath	15	Hinkle No.4	Jack B. Hinkle	Kurtis 4000	Offy	6	132	Flagged	446
George Connor	20	Ricketts No.90		Kurtis 4000	Offy	17	26	Brkn throttle shaft	223
Bob Sweikert	21	Tom Sarafoff No.22	Tom Sarafoff	Kurtis 2000	Offy	1	24	Spun out, turn 2	148
Eddie Russo	22	Pritchard No.19	Charles Pritchard	Kurtis 2000	Offy	21	19	Stalled engine	148
Jim Rathmann	99	Belanger No.99	Murrell Belanger	Kurtis	Offy		0	DNS, too slow	0
Mike Nazaruk	99	McNamara No.73	Lee Elkins	Kurtis 2000	Offy		0	DNS, too slow	30
Allen Heath	99	John Zink No.66	M. A. Walker	Kurtis	Offy		0	DNS, too slow	30
Spider Webb	99	Rouge Motor No.32	John T. Stanko	Kurtis 4000	Offy		0	DNS, too slow	30
Andy Linden	99	Miracle Power No.9	Hart Fullerton	Kurtis 4000	OffySc	0		DNS, too slow	30
Detroit, MI — 100 Mile — 30 Aug									
Paul Russo	2	Belanger No.99	Murrell Belanger	Kurtis	Offy	4	99	Flagged	1,728
Chuck Stevenson	3	Springfield Welding No.16	Bessie Lee Paoli	Kurtis 4000	Offy	15	98	Flagged	960
Jack McGrath	4	Hinkle No.4	Jack B. Hinkle	Kurtis 4000	Offy	7	98	Flagged	624
Neal Carter	5	Auto Shippers No.25	Eugene A. Casaroll	Kurtis 500A	Offy	13	98	Flagged	576
Andy Linden	8	Miracle Power No.9	Hart Fullerton	Kurtis 4000	OffySc	2	96	Flagged	360
Bobby Ball	9	Blakely Oil No.15	John McDaniel	Kurtis 4000	Offy	5	96	Flagged	336
Walt Faulkner	12	Lutes Truck Parts No.10	W.J. Lutes	Kurtis 4000	Offy	18	51	Wrecked, turn 4 flip	240
Jimmy Reece	15	John Zink No.44	John Ziink	Kurtis 4000	Offy	1	34	Bent underpan	168
Rodger Ward	16	Federal Engineering No.34	Federal Automotive Assoc.	Kurtis 4000	Offy	3	19	Brkn connecting rod	144
Mike Nazaruk	17	McNamara No.73	Lee Elkins	Kurtis 2000	Offy	6	15	Broken drive shaft	96
Joe James	98	Lutes Truck Parts No.10	W.J. Lutes	Kurtis 4000	Offy		35	Rel Faulkner 52-86	0
Chuck Weyant	99	Mel-Rae No.67	Mel B. Wiggers	Kurtis 4000	Offy		0	DNS, too slow	0
George Connor	99	Ricketts No.90		Kurtis 4000	Offy		0	DNS, too slow	0
Sam Hanks	99	Bardahl No.14	Ed Walsh	Kurtis 4000	Offy		0	DNS, too slow	0
Cliff Griffith	99	Dillon No.19		Kurtis 4000	Offy		0	DNS, too slow	0
Joe James	99	Bowes Seal Fast No.33	Ray T. Brady	Kurtis 4000	Offy		0	DNS, too slow	0
Frank Luptow	99	Walker No.66	M. A. Walker	Kurtis	Offy		0	DNS, too slow	0
Bob Sweikert	99	Tom Sarafoff No.22	Tom Sarafoff	Kurtis 2000	Offy		0	Did not qualify	0
DuQuoin, IL — 100 Mile — 1 Sep									
Chuck Stevenson	1	Springfield Welding No.16	Bessie Lee Paoli	Kurtis 4000	Offy	6	100	1:07:52.00 88.409	2,950
Sam Hanks	2	Bardahl No.14	Ed Walsh	Kurtis 4000	Offy	11	100	Finished	2,124
Paul Russo	4	Belanger No.99	Murrell Belanger	Kurtis	Offy	3	100	Finished	767
Jack McGrath	5	Hinkle No.4	Jack B. Hinkle	Kurtis 4000	Offy	4	100	Finished	708
Bobby Ball	6	Blakely Oil No.15	John McDaniel	Kurtis 4000	Offy	8	100	Finished	590
Joe James	9	Bowes Seal Fast No.33	Ray T. Brady	Kurtis 4000	Offy	5	100	Finished	413
Cliff Griffith	11	Dillon No.19		Kurtis 4000	Offy	15	98	Flagged	354
Bob Sweikert	12	Tom Sarafoff No.22	Tom Sarafoff	Kurtis 2000	Offy	2	98	Flagged	295
Jimmy Reece	15	John Zink No.44	John Ziink	Kurtis 4000	Offy	14	97	Flagged	206
Chuck Weyant	16	Mel-Rae No.67	Mel B. Wiggers	Kurtis 4000	Offy	18	95	Flagged	177
George Connor	99	Ricketts No.90		Kurtis 4000	Offy		0	DNS, too slow	0
Andy Linden	99	Miracle Power No.9	Hart Fullerton	Kurtis 4000	OffySc		0	DNS, too slow	0
Neal Carter	99	Auto Shippers No.25	Eugene A. Casaroll	Kurtis 500A	Offy		0	DNS, too slow	0
George Tichenor	99	Dr. Morris No.93	Ludson D. Morris, MD	Kurtis 2000	Offy		0	DNS, too slow	0
Buzz Barton	99	Walker No.66	M. A. Walker	Kurtis	Offy		0	DNS, too slow	0
Syracuse, NY — 100 Mile — 6 Sep									
Jack McGrath	1	Hinkle No.4	Jack B. Hinkle	Kurtis 4000	Offy	1	100	1:07:21.64 89.224	3,750
Mike Nazaruk	2	McNamara No.73	Lee Elkins	Kurtis 2000	Offy	2	100	Finished	2,700
Bobby Ball	3	Blakely Oil No.15	John McDaniel	Kurtis 4000	Offy	16	100	Finished	1,500
Paul Russo	4	Belanger No.99	Murrell Belanger	Kurtis	Offy	6	100	Finished	975
Jimmy Reece	5	John Zink No.44	John Ziink	Kurtis 4000	Offy	5	100	Finished	900
Chuck Stevenson	7	Springfield Welding No.16	Bessie Lee Paoli	Kurtis 4000	Offy	17	100	Finished	600
Joe James	9	Bowes Seal Fast No.33	Ray T. Brady	Kurtis 4000	Offy	9	100	Finished	525
Sam Hanks	12	Bardahl No.14	Ed Walsh	Kurtis 4000	Offy	7	100	Finished	375
Rodger Ward	13	Federal Engineering No.34	Federal Automotive Associates	Kurtis 4000	Offy	10	100	Finished	337
Neal Carter	99	Auto Shippers No.25	Eugene A. Casaroll	Kurtis 500A	Offy		0	DNS, too slow	0
Andy Linden	99	Miracle Power No.9	Hart Fullerton	Kurtis 4000	OffySc	0		DNS, too slow	0
Denver, CO — 100 Mile — 28 Sep									
Chuck Stevenson	2	Springfield Welding No.16	Bessie Lee Paoli	Kurtis 4000	Offy	7	100	Finished	1,869
Bobby Ball	3	Blakely Oil No.15	John McDaniel	Kurtis 4000	Offy	6	100	Finished	1,038
Sam Hanks	4	Bardahl No.14	Ed Walsh	Kurtis 4000	Offy	15	100	Finished	675

Driver	Fin Pos	Car Name	Entrant	Chassis	Engine	Start Pos	Laps	Time/Speed/Out	Purse
Jack McGrath	6	Hinkle No.4	Jack B. Hinkle	Kurtis 4000	Offy	3	99	Flagged	519
Neal Carter	7	Lutes Truck Parts No.10	W.J. Lutes	Kurtis 4000	Offy	2	99	Flagged	415
Jimmy Reece	10	John Zink No.44	John Ziink	Kurtis 4000	Offy	13	99	Flagged	337
Mike Nazaruk	12	McNamara No.73	Lee Elkins	Kurtis 2000	Offy	16	98	Flagged	259
Danny Oakes	13	Ricketts No.90		Kurtis 4000	Offy	10	96	Flagged	233
Bob Scott	15	Federal Engineering No.34	Federal Automotive Assoc.	Kurtis 4000	Offy	14	96	Flagged	181
Red Hamilton	16	Dillon No.19		Kurtis 4000	Offy	5	95	Flagged	155
Paul Russo	18	Belanger No.99	Murrell Belanger	Kurtis	Offy	11	49	Stalled BS, magneto	103
Rodger Ward	99	John Zink No.66	M. A. Walker	Kurtis	Offy		0	DNS, too slow	0
Andy Linden	99	Miracle Power No.9	Hart Fullerton	Kurtis 4000	OffySc	0		DNS, too slow	0
San Jose, CA — 100 Mile — 2 Nov									
Bobby Ball	1	Blakely Oil No.15	John McDaniel	Kurtis 4000	Offy	3	100	No time	1,875
Mike Nazaruk	2	McNamara No.73	Lee Elkins	Kurtis 2000	Offy	5	100	Finished	1,350
Paul Russo	3	Belanger No.99	Murrell Belanger	Kurtis	Offy	10	100	Finished	750
Jack McGrath	4	Hinkle No.4	Jack B. Hinkle	Kurtis 4000	Offy	1	99	Flagged	488
Sam Hanks	5	Bardahl No.14	Ed Walsh	Kurtis 4000	Offy	15	99	Flagged	450
Bob Scott	6	Federal Engineering No.34	Federal Automotive Assoc.	Kurtis 4000	Offy	12	98	Flagged	375
Chuck Stevenson	7	Springfield Welding No.16	Bessie Lee Paoli	Kurtis 4000	Offy	14	98	Flagged	300
Jimmy Reece	10	John Zink No.44	John Ziink	Kurtis 4000	Offy	9	32	Flagged	244
Andy Linden	15	Miracle Power No.9	Hart Fullerton	Kurtis 4000	OffySc	6	38	Magneto	131
Johnnie Parsons	16	Ricketts No.6		Kurtis 4000	Offy	7	37	Engine overheated	113
Rodger Ward	18	Dobry No.63	Leo Dobry	Kurtis 2000	Offy	11	8	Broken gear	75
Duane Carter	98	John Zink No.44	John Ziink	Kurtis 4000	Offy		64	Rel Reece 33-96	0
Cal Niday	99	Lutes Truck Parts No.10	W.J. Lutes	Kurtis 4000	Offy		0	DNS, too slow	0
Phoenix, AZ — 100 Mile — 11 Nov									
Johnnie Parsons	1	Ricketts No.6		Kurtis 4000	Offy	2	100	1:09:52.00 85.87	2,851
Sam Hanks	2	Bardahl No.14	Ed Walsh	Kurtis 4000	Offy	8	100	Finished	2,052
Bobby Ball	3	Blakely Oil No.15	John McDaniel	Kurtis 4000	Offy	7	99	Flagged	1,140
Manuel Ayulo	4	Lutes Truck Parts No.10	W.J. Lutes	Kurtis 4000	Offy	4	99	Flagged	741
Chuck Stevenson	6	Springfield Welding No.16	Bessie Lee Paoli	Kurtis 4000	Offy	13	99	Flagged	570
Andy Linden	7	Miracle Power No.9	Hart Fullerton	Kurtis 4000	OffySc	11	99	Flagged	456
Rodger Ward	9	Walker No.66	M. A. Walker	Kurtis	Offy	9	98	Flagged	399
Jimmy Reece	11	John Zink No.44	John Ziink	Kurtis 4000	Offy	14	96	Flagged	342
Paul Russo	15	Belanger No.99	Murrell Belanger	Kurtis	Offy	15	42	Magneto	200
Jack McGrath	16	Hinkle No.4	Jack B. Hinkle	Kurtis 4000	Offy	1	35	Lost wheel	171
Red Hamilton	17	Dillon No.19		Kurtis 4000	Offy	16	33	Gear trouble	114
Mike Nazaruk	18	McNamara No.73	Lee Elkins	Kurtis 2000	Offy	6	19	Gear trouble	114
Bob Scott	99	Federal Engineering No.34	Federal Automotive Assoc.	Kurtis 4000	Offy		0	Did not qualify	0
Williams Grove, PA — 25 Mile — 27 Jul									
Paul Russo	1	Lutes Truck Parts No.10	W.J. Lutes	Kurtis 4000	Offy	1	50	0:22:35.53 66.395	1,237
Johnnie Parsons	6	Ricketts No.6		Kurtis 4000	Offy	6	50	Finished	337
Joe James	7	Bowes Seal Fast No.33	Ray T. Brady	Kurtis 4000	Offy	11	50	Finished	525
Neal Carter	8	Auto Shippers No.25	Eugene A. Casaroll	Kurtis 500A	Offy	14	50	Finished	337
Sam Hanks	11	Bardahl No.14	Ed Walsh	Kurtis 4000	Offy	3	45	Spun out, turn 2	262
Bob Scott	99	Dr. Morris No.93	Ludson D. Morris, MD	Kurtis 2000	Offy	19	0	DNS, too slow	150
Pikes Peak, CO — 12.42 Mile — 1 Sep									
George Hammond	1	Dobry No.63	Leo Dobry	Kurtis	Offy	6	1	0:15:30.69 48.042	3,500
1953									
Indianapolis, IN — 500 Mile — 30 May									
Bill Vukovich	1	Fuel Injection No.14	Howard Keck Co.	Kurtis 500A	Offy	1	200	3:53:01.69 128.740	89,496
Art Cross	2	Springfield Welding No.16	Bessie Lee Paoli	Kurtis 4000	Offy	12	200	3:56:32.56 126.827	27,296
Sam Hanks	3	Bardahl No.3	Ed Walsh	Kurtis 4000	Offy	9	151	3:57:13.24 126.465	16,421
Fred Agabashian	4	Elgin Piston Pin/Grancor No.59	Grancor Auto Specialists	Kurtis 500B	Offy	2	104	3:57:40.91 126.465	12,946
Jack McGrath	5	Hinkle No.5	Jack B. Hinkle	Kurtis 4000	Offy	3	200	4:00:51.33 124.556	10,621
Jimmy Daywalt	6	Sumar No.48	Chapman S. Root	Kurtis 3000	Offy	21	200	4:01:11.88 124.379	7,696
Jim Rathmann	7	Travelon Trailer No.2	Ernest L. Ruiz	Kurtis 500B	Offy	25	112	4:01:47.65 124.072	6,846
Jimmy Davies	10	Pat Clancy No.53	Pat Clancy	Kurtis 500B	Offy	32	193	Flagged 119.203	5,546
Duke Nalon	11	Novi Governor No.9	Jean Marsenac	Kurtis FD	NoviSc	26	191	Spun out NE turn	3,316
Carl Scarborough	12	McNamara No.73	Lee Elkins	Kurt/WettD	Offy	19	69	Flagged	3,146
Bill Holland	15	Crawford No.49	Ray Crawford	Kurtis 500B	Offy	28	141	Broken cam gear	3,236
Rodger Ward	16	M. A. Walker No.92	M. A. Walker	Kurtis D	Offy	10	138	Stalled	2,916
Walt Faulkner	17	Auto Shippers No.23	Eugene A. Casaroll	Kurtis 500A	Offy	14	134	Flagged	2,496
Marshall Teague	18	Hart Fullerton\Pure Oil No.22	Hart Fullerton	Kurtis 4000	Offy	22	169	Oil leak	2,376
Spider Webb	19	Lubri-Loy No.62	3-L Racing Team	Kurtis 3000	Offy	18	112	Oil leak	2,346
Pat Flaherty	22	Peter Schmidt No.77	Peter Schmidt	Kurtis 3000	Offy	24	115	Wrecked NS	2,256
Jerry Hoyt	23	John Zink No.55	John Zink	Kurtis 4000	Offy	7	82	Overheated	2,226
Paul Russo	25	Federal Engineering No.7	Federal Automotive Assoc.	Kurtis 3000	Offy	17	89	Magneto	2,166
Johnnie Parsons	26	Belond Equa Flow No.21	J. S. Belond	Kurtis 500B	Offy	8	86	Broken crankshaft	2,636
Gene Hartley	28	Federal Engineering No.41	Federal Automotive Assoc.	Kurtis 4000	Offy	13	53	Wrecked NE	2,076
Cal Niday	30	Miracle Power No.99	Murrell Belanger	Kurtis D	Offy	30	30	Magneto	2,316

Driver	Fin Pos	Car Name	Entrant	Chassis	Engine	Start Pos	Laps	Time/Speed/Out	Purse
Eddie Johnson	98	Travelon Trailer No.2	Ernest L. Ruiz	Kurtis 500B	Offy		88	Rel J Rathmn 113-200	0
Duane Carter	98	Bardahl No.3	Ed Walsh	Kurtis 4000	Offy		49	Rel Hanks 152-200	0
Johnny Mantz	98	Auto Shippers No.23	Eugene A. Casaroll	Kurtis 500A	Offy		42	Rel Faulkner 135-176	0
Jim Rathmann	98	Crawford No.49	Ray Crawford	Kurtis 500B	Offy		36	Rel Holland 142-177	0
Chuck Stevenson	98	John Zink No.55	John Zink	Kurtis 4000	Offy		13	Rel Hoyt 83-95	0
Andy Linden	98	John Zink No.55	John Zink	Kurtis 4000	Offy		12	Rel Hoyt 96-107	0
Paul Russo	98	Elgin Piston Pin/Grancor No.59	Grancor Auto Specialists	Kurtis 500B	Offy		96	Rel Agabashn 105-200	0
Jackie Holmes	98	Lubri-Loy No.62	3-L Racing Team	Kurtis 3000	Offy		9	Rel Webb 158-166	0
Johnny Thomson	98	Lubri-Loy No.62	3-L Racing Team	Kurtis 3000	Offy		45	Rel Webb 113-157	0
Bob Scott	98	McNamara No.73	Lee Elkins	Kurt/WettD	Offy		121	Rel S'borough 70-190	0
Duke Dinsmore	98	M. A. Walker No.92	M. A. Walker	Kurtis D	Offy		10	Rel Ward 145-154	0
Andy Linden	98	M. A. Walker No.92	M. A. Walker	Kurtis D	Offy		29	Rel Ward 116-144	0
Eddie Johnson	99	City of Detroit No.26	W. J. Lutes	Kurtis 4000	Offy		0	DNS, too slow	0
Cal Niday	99	Storey-Ricketts No.78	Storey & Ricketts	Kurtis 4000	Offy		0	DNS, too slow	0
Johnny Tolan	99	Blakely Oil No.85	John T. McDaniel	Kurtis 4000	Offy		0	DNS, too slow	0
Bill Holland	99	Slick Racers No.61	Slick Racers, Inc	Kurtis 3000	Offy		0	DNS, too slow	0
Pat O'Connor	99	Engle-Stanko No.64	Rouge Motors, Inc	Kurtis 4000	Offy		0	DNS, too slow	0
Chet Miller	99	Novi Governor No.15	Jean Marcenac	Kurtis FD	NoviSc	0		Wr pract, SW, fatal	0
Joe Sostilio	99	Belanger Motors No.17	Murrell Belanger	Kurtis 500A	ChryV8	0		Did not qualify	0
Cliff Griffith	99	Bardahl No.24	Ed Walsh	Kurtis 500B	Offy		0	Wrecked practice, SW	0
George Connor	99	Chrysler\Wolcott No.25	Roger Wolcott	Kurtis 500A	ChryV8	0		Did not qualify	0
Art Cross	99	Ray Brady No.33	Ray Brady	Kurtis 4000	Offy		0	Did not qualify	0
Eddie Sachs	99	Dr. Morris No.34	Ludson D. Morris, MD	Kurtis 2000	Offy		0	Did not qualify	0
LeRoy Warriner	99	Jim Robbins No.44	Jim Robbins	Kurtis D	Offy		0	Did not qualify	0
Johnny Fedricks	99	Dunn Engineering No.46	Harry Dunn	Kurtis 500B	Offy		0	Did not qualify	0
Duke Dinsmore	99	Sarafoff No.52	Tom Sarafoff	Kurtis 2000	Offy		0	Did not qualify	0
Johnny Tolan	99	Chrysler\Wolcott No.66	Roger Wolcott	Kurtis 500A	ChryV8	0		Did not qualify	0
Jackie Holmes	99	Merkler Machine Works No.71	Franklin Merkler	Kurtis	OffySc	0		Did not qualify	0
Frank Armi	99	Mel-Rae No.79	Mel B. Wiggers	Kurtis 4000	Offy		0	Did not qualify	0
Johnny Roberts	99	Burns No.82	William P. Burns	Kurtis 2000	Offy		0	Did not qualify	0
Bill Boyd	99	Merz Engineering No.86	Merz Engineering, Inc	Kurtis 2000	Offy		0	Did not qualify	0
Bill Homeier	99	Cal Connell No.87	Calvin C. Connell	Kurtis	CadlV8	0		Did not qualify	0
Red Hamilton	99	Dillon No.91	Clyde R. Dillon	Kurtis 4000	Offy		0	Did not qualify	0
Al Herman	99	Traylor Engineering No.93	Samual W. Traylor IV	Kurtis 2000	Offy		0	Did not qualify	0
Milwaukee, WI — 100 Mile — 7 Jun									
Jack McGrath	1	Hinkle No.5	Jack B. Hinkle	Kurtis 4000	Offy	4	100	1:04:04.759 93.634	5,150
Jimmy Bryan	2	Springfield Welding No.16	Bessie Lee Paoli	Kurtis 4000	Offy	9	100	Finished	3,296
Jerry Hoyt	3	John Zink No.55	John Zink	Kurtis 4000	Offy	12	100	Finished	2,266
Cal Niday	6	Ricketts No.78		Kurtis 4000	Offy	3	100	Finished	1,030
Jimmy Davies	8	Belanger No.99	Murrell Belanger	Kurtis	Offy	14	99	Flagged	824
Sam Hanks	10	Bardahl No.3	Ed Walsh	Kurtis 4000	Offy	10	21	Flagged	618
Rodger Ward	18	M. A. Walker No.92	M. A. Walker	Kurtis	Offy	2	14	Blown engine	206
Bob Scott	98	Bardahl No.3	Ed Walsh	Kurtis 4000	Offy		77	Rel Hanks 22-98	0
Bill Homeier	99	Dr. Morris No.34	Ludson D. Morris, MD	Kurtis 2000	Offy		0	DNS, too slow	0
Walt Faulkner	99	Hart Fullerton\Pure Oil No.22	Hart Fullerton	Kurtis 4000	Offy		0	DNS, too slow	0
Chuck Marshall	99	Mel-Rae No.79	Mel B. Wiggers	Kurtis 4000	Offy		0	DNS, too slow	0
Eddie Johnson	99	City of Detroit No.27	W. J. Lutes	Kurtis 4000	Offy		0	DNS, too slow	0
Pat O'Connor	99	Peter Wales No.68	Peter Wales	Kurtis 4000	Offy		0	DNS, too slow	0
Johnny Roberts	99	Burns No.82	William P. Burns	Kurtis 2000	Offy		0	DNS, too slow	0
Johnny Fedricks	99	Dunn Engineering No.89	Harry Dunn	Kurtis 500B	Offy		0	DNS, too slow	0
Springfield, IL — 100 Mile — 21 Jun									
Rodger Ward	1	M. A. Walker No.92	M. A. Walker	Kurtis	Offy	3	100	1:07:03.90 89.465	1,875
Sam Hanks	2	Bardahl No.3	Ed Walsh	Kurtis 4000	Offy	6	100	Finished	1,350
Jack McGrath	4	Hinkle No.5	Jack B. Hinkle	Kurtis 4000	Offy	2	100	Finished	487
Jimmy Bryan	8	Springfield Welding No.16	Bessie Lee Paoli	Kurtis 4000	Offy	11	100	Finished	281
Walt Faulkner	9	Hart Fullerton\Pure Oil No.22	Hart Fullerton	Kurtis 4000	Offy	17	98	Flagged	262
Bill Homeier	12	Dr. Morris No.93	Ludson D. Morris, MD	Kurtis 2000	Offy	8	92	Flagged	187
Chuck Marshall	13	Mel-Rae No.79	Mel B. Wiggers	Kurtis 4000	Offy	16	91	Flagged	112
Jerry Hoyt	14	John Zink No.55	John Zink	Kurtis 4000	Offy	7	76	Wrecked, turn 3	150
Allen Heath	15	McNamara No.73	Lee Elkins	Kurt/WettD	Offy	14	49	Out	131
Gene Hartley	99	Burns No.82	William P. Burns	Kurtis 2000	Offy		0	DNS, too slow	0
Eddie Johnson	99	City of Detroit No.27	W. J. Lutes	Kurtis 4000	Offy		0	DNS, too slow	0
Johnny Fedricks	99	Dunn Engineering No.89	Harry Dunn	Kurtis 500B	Offy		0	DNS, too slow	0
Duke Dinsmore	99	Sarafoff No.52	Tom Sarafoff	Kurtis 2000	Offy		0	DNS, too slow	0
Eddie Russo	99	Pete Wales No.68	Peter Wales	Kurtis 4000	Offy		0	DNS, too slow	0
Detroit, MI — 100 Mile — 4 Jul									
Rodger Ward	1	M. A. Walker No.92	M. A. Walker	Kurtis	Offy	3	51	0:37:12.04 82.257	1,875
Allen Heath	2	McNamara No.73	Lee Elkins	Kurt/Wett	Offy	10	51	Finished	1,350
Jimmy Bryan	3	Springfield Welding No.16	Bessie Lee Paoli	Kurtis 4000	Offy	12	51	Finished	750
Sam Hanks	4	Bardahl No.3	Ed Walsh	Kurtis 4000	Offy	13	49	Flagged	487
Johnnie Parsons	8	Belond Equa Flow No.21	Louis & Bruce Bromme	Kurtis 4000	Offy	4	48	Brkn shocks	281
Duke Dinsmore	12	Sarafoff No.52	Tom Sarafoff	Kurtis 2000	Offy	6	47	Flagged	187

Kurtis-Kraft

Driver	Fin Pos	Car Name	Entrant	Chassis	Engine	Start Pos	Laps	Time/Speed/Out	Purse
Paul Russo	13	Federal Engineering No.7	Federal Automotive Assoc.	Kurtis 4000	Offy	9	46	Withdrew, ruf trac	169
Jack McGrath	16	Hinkle No.5	Jack B. Hinkle	Kurtis 4000	Offy	14	38	Withdrew, ruf track	112
Johnny Fedricks	18	Dunn Engineering No.89	Harry Dunn	Kurtis 500B	Offy	2	9	Out	75
Eddie Johnson	99	City of Detroit No.27	W. J. Lutes	Kurtis 4000	Offy		0	DNS, too slow	0
Johnny Roberts	99	Burns No.82	William P. Burns	Kurtis 2000	Offy		0	DNS, too slow	0
Walt Faulkner	99	Pete Wales No.68	Peter Wales	Kurtis 4000	Offy		0	DNS, too slow	0
Bill Homeier	99	Dr. Morris No.93	Ludson D. Morris, MD	Kurtis 2000	Offy		0	DNS, too slow	0
Springfield, IL — 100 Mile — 22 Aug									
Sam Hanks	1	Bardahl No.3	Ed Walsh	Kurtis 4000	Offy	5	100	1:05:49.00 91.162	4,599
Johnnie Parsons	4	Belond Equa Flow No.21	Louis & Bruce Bromme	Kurtis 4000	Offy	16	100	Finished	1,472
Cal Niday	5	Belanger No.99	Murrell Belanger	Kurtis	Offy	3	100	Finished	1,104
Paul Russo	6	Federal Engineering No.7	Federal Automotive Assoc.	Kurtis 4000	Offy	6	100	Finished	920
Jimmy Reece	8	Pete Wales No.68	Peter Wales	Kurtis 4000	Offy	12	100	Finished	736
Jack McGrath	13	Hinkle No.5	Jack B. Hinkle	Kurtis 4000	Offy	13	87	Flagged	368
Johnny Fedricks	14	Dunn Engineering No.89	Harry Dunn	Kurtis 500B	Offy	15	75	Flagged	368
Jimmy Bryan	15	Springfield Welding No.16	Bessie Lee Paoli	Kurtis 4000	Offy	14	59	Spun out, turn 3	368
Rodger Ward	17	M. A. Walker No.92	M. A. Walker	Kurtis	Offy	9	33	Connect rod bearing	184
Walt Faulkner	18	Auto Shippers No.23	Eugene A. Casaroll	Kurtis 4000	Offy	17	14	Brkn radiator hose	184
Eddie Johnson	99	Dr. Morris No.93	Ludson D. Morris, MD	Kurtis 2000	Offy		0	DNS, too slow	0
Ralph Pratt	99	City of Detroit No.27	W. J. Lutes	Kurtis 4000	Offy		0	DNS, too slow	0
Bill Holland	99	McNamara No.73	Lee Elkins	Kurt/Wett	Offy		0	DNS, too slow	0
Jerry Hoyt	99	John Zink No.55	John Zink	Kurtis 4000	Offy		0	DNS, too slow	0
Johnny Roberts	99	Burns No.82	William P. Burns	Kurtis 2000	Offy		0	DNS, too slow	0
George Tichenor	99	Belanger No.61	Murrell Belanger	Kurtis 500A	ChryV8	0		DNS, too slow	0
Chuck Marshall	99	Mel-Rae No.79	Mel B. Wiggers	Kurtis 4000	Offy		0	Did not qualify	0
Milwaukee, WI — 200 Mile — 30 Aug									
Paul Russo	3	Federal Engineering No.7	Federal Automotive Assoc.	Kurtis 4000	Offy	21	200	Finished	2,650
Jack McGrath	4	Hinkle No.5	Jack B. Hinkle	Kurtis 4000	Offy	1	200	Finished	1,988
Rodger Ward	6	M. A. Walker No.92	M. A. Walker	Kurtis	Offy	10	200	Finished	1,325
Jimmy Reece	7	Pete Wales No.68	Peter Wales	Kurtis 4000	Offy	19	198	Flagged	1,060
Jimmy Bryan	10	Springfield Welding No.16	Bessie Lee Paoli	Kurtis 4000	Offy	8	198	Flagged	861
Duke Dinsmore	15	Sarafoff No.52	Tom Sarafoff	Kurtis 2000	Offy	20	188	Flagged	397
Bill Holland	18	McNamara No.73	Lee Elkins	Kurt/Wett	Offy	7	121	Flagged	265
Cal Niday	19	Belanger No.99	Murrell Belanger	Kurtis	Offy	16	170	Brkn differential	199
Jerry Hoyt	20	John Zink No.55	John Zink	Kurtis 4000	Offy	22	101	Flipped, turn 3	199
Johnnie Parsons	21	Belond Equa Flow No.21	Louis & Bruce Bromme	Kurtis 4000	Offy	4	38	Broken oil line	132
Sam Hanks	22	Bardahl No.3	Ed Walsh	Kurtis 4000	Offy	5	23	Engine	132
Tony Bettenhausen	98	Belond Equa Flow No.21	Louis & Bruce Bromme	Kurtis 4000	Offy		5	Rel Parsons 39-43	0
Tony Bettenhausen	98	McNamara No.73	Lee Elkins	Kurt/Wett	Offy		62	Rel Holland 122-183	0
Johnny Roberts	99	Burns No.82	William P. Burns	Kurtis 2000	Offy		0	DNS, too slow	0
Chuck Weyant	99	Dr. Morris No.93	Ludson D. Morris, MD	Kurtis 2000	Offy		0	DNS, too slow	0
Walt Faulkner	99	Auto Shippers No.23	Eugene A. Casaroll	Kurtis 4000	Offy		0	DNS, too slow	0
Ralph Pratt	99	City of Detroit No.27	W. J. Lutes	Kurtis 4000	Offy		0	DNS, too slow	0
Johnny Fedricks	99	Dunn Engineering No.89	Harry Dunn	Kurtis 500B	Offy		0	DNS, too slow	0
DuQuoin, IL — 100 Mile — 7 Sep									
Sam Hanks	1	Bardahl No.3	Ed Walsh	Kurtis 4000	Offy	3	100	1:06:40.71 89.984	3,762
Johnnie Parsons	3	Belond Equa Flow No.21	Louis & Bruce Bromme	Kurtis 4000	Offy	11	100	Finished	1,505
Jack McGrath	4	Hinkle No.5	Jack B. Hinkle	Kurtis 4000	Offy	8	100	Finished	978
Walt Faulkner	6	Auto Shippers No.23	Eugene A. Casaroll	Kurtis 4000	Offy	13	100	Finished	752
Jimmy Bryan	7	Springfield Welding No.16	Bessie Lee Paoli	Kurtis 4000	Offy	15	98	Flagged	602
Mike Nazaruk	13	McNamara No.73	Lee Elkins	Kurt/Wett	Offy	1	98	Flagged	339
Paul Russo	14	Federal Engineering No.7	Federal Automotive Assoc.	Kurtis 4000	Offy	17	98	Flagged	301
Johnny Roberts	16	Burns No.82	William P. Burns	Kurtis 2000	Offy	12	96	Flagged	226
Tony Bettenhausen	18	Belanger No.99	Murrell Belanger	Kurtis	Offy	6	85	Oil leak	150
Johnny Fedricks	99	Dunn Engineering No.89	Harry Dunn	Kurtis 500B	Offy		0	DNS, too slow	0
Jimmy Reece	99	Pete Wales No.68	Peter Wales	Kurtis 4000	Offy		0	DNS, too slow	0
Jerry Hoyt	99	Dr. Morris No.93	Ludson D. Morris, MD	Kurtis 2000	Offy		0	DNS, too slow	0
Rodger Ward	99	M. A. Walker No.92	M. A. Walker	Kurtis	Offy		0	DNS, too slow	0
Chuck Marshall	99	Mel-Rae No.79	Mel B. Wiggers	Kurtis 4000	Offy		0	DNS, too slow	0
Bob Scott	99	City of Detroit No.27	W. J. Lutes	Kurtis 4000	Offy		0	DNS, too slow	0
Johnny Hobel	99	Shouse Motors No.61		Kurtis 500A	ChryV8	0		DNS, too slow	0
Pikes Peak, CO — 12 Mile — 7 Sep									
Louis Unser	1	Federal Engineering No.7	Federal Automotive Assoc.	Kurtis 4000	Offy	13	1	0:15:15.40 48.844	4,000
Syracuse, NY — 100 Mile — 12 Sep									
Tony Bettenhausen	1	Belanger No.99	Murrell Belanger	Kurtis	Offy	5	100	1:20:59 74.089	3,750
Mike Nazaruk	2	McNamara No.73	Lee Elkins	Kurt/Wett	Offy	2	100	Finished	2,700
Sam Hanks	4	Bardahl No.3	Ed Walsh	Kurtis 4000	Offy	15	99	Flagged	975
Jerry Hoyt	6	Dr. Morris No.93	Ludson D. Morris, MD	Kurtis 2000	Offy	8	97	Flagged	750
Rodger Ward	7	M. A. Walker No.92	M. A. Walker	Kurtis	Offy	7	47	Flagged	600
Jack McGrath	9	Hinkle No.5	Jack B. Hinkle	Kurtis 4000	Offy	11	71	Wrecked, turn 3	525
Johnnie Parsons	12	Belond Equa Flow No.21	Louis & Bruce Bromme	Kurtis 4000	Offy	14	64	Low oil pressure	375

Driver	Fin Pos	Car Name	Entrant	Chassis	Engine	Start Pos	Laps	Time/Speed/Out	Purse
Paul Russo	13	Federal Engineering No.7	Federal Automotive Assoc.	Kurtis 4000	Offy	4	63	Withdrew	337
Johnny Fedricks	15	Dunn Engineering No.89	Harry Dunn	Kurtis 500B	Offy	17	28	Wrecked, turn 4	262
Jimmy Bryan	17	Springfield Welding No.16	Bessie Lee Paoli	Kurtis 4000	Offy	6	0	Wrecked, turn 3	150
Bob Scott	98	M. A. Walker No.92	M. A. Walker	Kurtis	Offy		50	Rel Ward 48-97	0
Walt Faulkner	99	Auto Shippers No.23	Eugene A. Casaroll	Kurtis 4000	Offy		0	DNS, too slow	0
Jimmy Reece	99	Pete Wales No.68	Peter Wales	Kurtis 4000	Offy		0	DNS, too slow	0
Johnny Roberts	99	Burns No.82	William P. Burns	Kurtis 2000	Offy		0	DNS, too slow	0
Len Duncan	99	Ray Brady No.33	Ray Brady	Kurtis 4000	Offy		0	DNS, too slow	0
Indianapolis, IN — 100 Mile — 26 Sep									
Johnnie Parsons	3	Belond Equa Flow No.21	Louis & Bruce Bromme	Kurtis 4000	Offy	5	100	Finished	2,508
Sam Hanks	5	Bardahl No.3	Ed Walsh	Kurtis 4000	Offy	6	100	Finished	1,501
Tony Bettenhausen	6	Belanger No.99	Murrell Belanger	Kurtis	Offy	8	100	Finished	1,164
Rodger Ward	8	M. A. Walker No.92	M. A. Walker	Kurtis	Offy	16	100	Finished	863
Jimmy Bryan	11	Springfield Welding No.16	Bessie Lee Paoli	Kurtis 4000	Offy	7	100	Finished	672
Paul Russo	12	Federal Engineering No.7	Federal Automotive Assoc.	Kurtis 4000	Offy	15	100	Finished	552
Rex Easton	13	Mel-Rae No.79	Mel B. Wiggers	Kurtis 4000	Offy	9	100	Finished	487
Mike Nazaruk	15	McNamara No.73	Lee Elkins	Kurt/Wett	Offy	11	100	Finished	379
LeRoy Warriner	18	Lubri-Loy No.62	M. A. Walker	Kurtis 3000	Offy	18	38	Lost oil pressure	216
Walt Faulkner	99	Auto Shippers No.23	Eugene A. Casaroll	Kurtis 4000	Offy		0	DNS, too slow	0
Chuck Stevenson	99	John Zink No.55	John Zink	Kurtis 4000	Offy		0	DNS, too slow	0
Jimmy Reece	99	Pete Wales No.68	Peter Wales	Kurtis 4000	Offy		0	DNS, too slow	0
Johnny Fedricks	99	Dunn Engineering No.89	Harry Dunn	Kurtis 500B	Offy		0	DNS, too slow	0
Gene Force	99	Dr. Morris No.93	Ludson D. Morris, MD	Kurtis 2000	Offy		0	DNS, too slow	0
Johnny Roberts	99	Burns No.82	William P. Burns	Kurtis 2000	Offy		0	DNS, too slow	0
Potsy Goacher	99	Shouse Motors No.61		Kurtis 500A	ChryV8	0		DNS, too slow	0
Jack McGrath	99	Hinkle No.5	Jack B. Hinkle	Kurtis 4000	Offy		0	DNQ, burned piston	0
Sacramento, CA — 100 Mile — 25 Oct									
Jimmy Bryan	1	Springfield Welding No.16	Bessie Lee Paoli	Kurtis 4000	Offy	3	100	1:15:12.05 79.786	2,414
Sam Hanks	3	Bardahl No.3	Ed Walsh	Kurtis 4000	Offy	17	99	Flagged	968
Paul Russo	6	Federal Engineering No.7	Federal Automotive Assoc.	Kurtis 4000	Offy	11	96	Flagged	482
Jimmy Reece	7	Hart Fullerton\Pure Oil No.22	Hart Fullerton	Kurtis 4000	Offy	9	96	Flagged	386
Jack McGrath	10	Hinkle No.5	Jack B. Hinkle	Kurtis 4000	Offy	2	88	Flagged	313
Rodger Ward	13	M. A. Walker No.92	M. A. Walker	Kurtis	Offy	15	33	Engine	217
Mike Nazaruk	14	McNamara No.73	Lee Elkins	Kurt/WettD	Offy	1	13	Wrecked, turn 1	193
Johnnie Parsons	15	Belond Equa Flow No.21	Louis & Bruce Bromme	Kurtis 4000	Offy	5	13	Wrecked, turn 1	169
Don Olds	16	Leo Dobry No.63	Leo Dobry	Kurtis 2000	Offy	10	12	Wrecked, turn 1	144
Rex Easton	99	Mel Rae No.79	Mel B. Wiggers	Kurtis 4000	Offy		0	DNS, too slow	0
Bill Vukovich	99	Auto Shippers No.23	Eugene A. Casaroll	Kurtis 4000	Offy		0	DNS, too slow	0
Phoenix, AZ — 100 Mile — 11 Nov									
Tony Bettenhausen	1	Belanger No.99	Murrell Belanger	Kurtis	Offy	1	100	1:11:30.00 83.916	2,286
Duane Carter	2	Auto Shippers No.23	Eugene A. Casaroll	Kurtis 4000	Offy	15	99	Flagged	1,646
Sam Hanks	3	Bardahl No.3	Ed Walsh	Kurtis 4000	Offy	5	99	Flagged	914
Rex Easton	11	Mel-Rae No.79	Mel B. Wiggers	Kurtis 4000	Offy	11	92	Flagged	274
Jack McGrath	13	Hinkle No.5	Jack B. Hinkle	Kurtis 4000	Offy	16	75	Out	206
Jimmy Reece	14	Hart Fullerton\Pure Oil No.22	Hart Fullerton	Kurtis 4000	Offy	9	48	Out	183
Jimmy Bryan	15	Springfield Welding No.16	Bessie Lee Paoli	Kurtis 4000	Offy	18	44	Broken rear end	160
Johnnie Parsons	16	Belond Equa Flow No.21	Louis & Bruce Bromme	Kurtis 4000	Offy	6	42	Broken shocks	137
Paul Russo	17	Federal Engineering No.7	Federal Automotive Assoc.	Kurtis 4000	Offy	10	38	Broken suspension	91
Rodger Ward	18	M. A. Walker No.92	M. A. Walker	Kurtis	Offy	3	19	Broken suspension	91
Don Olds	99	Leo Dobry No.63	Leo Dobry	Kurtis 2000	Offy		0	Did not qualify	0
Eddie Sachs	99	McNamara No.73	Lee Elkins	Kurt/Wett	Offy		0	Flipped qualifying	0
Williams Grove, PA — 25 Mile — 26 Jul									
Johnnie Parsons	2	Belond Equa Flow No.21	Louis & Bruce Bromme	Kurtis 4000	Offy	6	50	Finished	862
Rodger Ward	3	M. A. Walker No.92	M. A. Walker	Kurtis	Offy	13	50	Finished	862
Paul Russo	4	Federal Engineering No.7	Federal Automotive Assoc.	Kurtis 4000	Offy	2	50	Finished	600
Johnny Roberts	10	Burns No.82	William P. Burns	Kurtis 2000	Offy	7	50	Finished	356
Ralph Pratt	12	City of Detroit No.27	W. J. Lutes	Kurtis 4000	Offy	12	49	Flagged	300
Allen Heath	13	McNamara No.73	Lee Elkins	Kurt/Wett	Offy	5	8	Broken driveshaft	225
Joe Sostilio	99	Dr. Morris No.93	Ludson D. Morris, MD	Kurtis 2000	Offy	17	0	DNS, too slow	75
Joe Barzda	99	Sarafoff No.52	Tom Sarafoff	Kurtis 2000	Offy	22	0	DNS, too slow	0
1954									
Indianapolis, IN — 500 Mile — 31 May									
Bill Vukovich	1	Fuel Injection No.14	Howard Keck Co.	Kurtis 500	Offy	19	200	3:49:17.27 130.840	74,934
Jack McGrath	3	Hinkle No.2	Jack B. Hinkle	Kurtis 500	Offy	1	200	3:50:36.97 130.086	27,404
Troy Ruttman	4	Auto Shippers No.34	Eugene A. Casaroll	Kurtis 500	Offy	11	130	3:52:09.90 129.218	12,709
Mike Nazaruk	5	McNamara No.73	Lee Elkins	Kurtis 500	Offy	14	200	3:52:41.85 128.923	10,934
Fred Agabashian	6	Merz Engineering No.77	Miklos Sperling	Kurtis 500	Offy	24	200	3:53:04.83 128.711	8,034
Paul Russo	8	Ansted Rotary No.5	Hoosier Racing Team, Inc	Kurtis 500	Offy	32	150	3:54:18.39 128.037	6,259
Larry Crockett	9	Federal Engineering No.28	Federal Automotive Assoc.	Kurtis 3000	Offy	25	200	3:56:24.56 126.899	6,984
Art Cross	11	Bardahl No.45	Ed Walsh	Kurtis 4000	Offy	27	120	3:57:39.42 126.232	5,254

Kurtis-Kraft

Driver	Fin Pos	Car Name	Entrant	Chassis	Engine	Start Pos	Laps	Time/Speed/Out	Purse
Manuel Ayulo	13	Peter Schmidt No.88	Peter Schmidt	Kurtis 500	Offy	22	197	Flagged	3,464
Bob Sweikert	14	Lutes Truck Parts No.17	Francis Bardazon	Kurtis 4000	Offy	9	197	Flagged	3,344
Duane Carter	15	Auto Shippers No.16	Eugene A. Casaroll	Kurtis 4000	Offy	8	76	Flagged	3,224
Ernie McCoy	16	Crawford No.32	Ray Crawford	Kurtis 500	Offy	20	194	Flagged	3,104
Sam Hanks	20	Bardahl No.1	Ed Walsh	Kurtis 4000	Offy	10	112	Wrecked, mainstretch	2,954
Pat O'Connor	21	Hopkins No.35	Motor Racers, Inc	Kurtis 500	Offy	12	181	Spun out, NE turn	3,274
Gene Hartley	23	John Zink No.31	John S. Zink	Kurtis 4000	Offy	17	151	Engine	2,814
Jerry Hoyt	26	Belanger No.99	Murrell Belanger	Kurtis D	Offy	30	130	Engine	2,624
Jimmy Daywalt	27	Sumar No.19	Chapman S. Root	Kurtis 500	Offy	2	111	Wrecked NW turn	4,444
Jim Rathmann	28	Bardahl No.38	Ed Walsh	Kurtis 500	Offy	28	95	Wrecked NW turn	2,764
Tony Bettenhausen	29	Mel Wiggers No.10	Mel B. Wiggers	Kurtis 500	Offy	21	105	Broken oil line	2,534
Johnnie Parsons	32	Belond Equa Flow No.15	So. Calif. Muffler Corp.	Kurtis 500	Offy	15	79	Broken piston	2,744
Bill Homeier	33	Jones & Maley No.51	Cars, Inc.	Kurtis 500	Offy	18	74	Hit pitwall	2,414
Jimmy Davies	98	Bardahl No.1	Ed Walsh	Kurtis 4000	Offy		36	Rel Hanks 113-148	0
Jim Rathmann	98	Bardahl No.1	Ed Walsh	Kurtis 4000	Offy		43	Rel Hanks 149-191	0
Jerry Hoyt	98	Ansted Rotary No.5	Hoosier Racing Team, Inc	Kurtis 500	Offy		50	Rel Russo 151-200	0
Marshall Teague	98	Auto Shippers No.16	Eugene A. Casaroll	Kurtis 4000	Offy		29	Rel Carter 77-105	0
Jimmy Jackson	98	Auto Shippers No.16	Eugene A. Casaroll	Kurtis 4000	Offy		15	Rel Carter 106-120	0
Tony Bettenhausen	98	Auto Shippers No.16	Eugene A. Casaroll	Kurtis 4000	Offy		34	Rel Carter 121-154	0
Jimmy Jackson	98	Auto Shippers No.16	Eugene A. Casaroll	Kurtis 4000	Offy		42	Rel Carter 155-196	0
Marshall Teague	98	John Zink No.31	John S. Zink	Kurtis 4000	Offy		17	Rel Hartley 152-168	0
Duane Carter	98	Auto Shippers No.34	Eugene A. Casaroll	Kurtis 500	Offy		70	Rel Ruttman 101-170	0
Pat Flaherty	98	Bardahl No.38	Ed Walsh	Kurtis 500	Offy		15	Rel JRathmann 96-110	0
Johnnie Parsons	98	Bardahl No.45	Ed Walsh	Kurtis 4000	Offy		22	Rel Cross 121-142	0
Sam Hanks	98	Bardahl No.45	Ed Walsh	Kurtis 4000	Offy		11	Rel Cross 143-153	0
Jimmy Davies	98	Bardahl No.45	Ed Walsh	Kurtis 4000	Offy		30	Rel Cross 171-200	0
Andy Linden	98	Bardahl No.45	Ed Walsh	Kurtis 4000	Offy		17	Rel Cross 154-170	0
Jimmy Davies	99	Pat Clancy No.53	Pat Clancy	Kurtis 500	Offy		0	DNS, too slow	0
Marshall Teague	99	Hart Fullerton No.3	Hart Fullerton	Kurtis 4000	Offy		0	DNS, too slow	0
Bob Scott	99	Travelon Trailer No.21	Ernest L. Ruiz	Kurtis 500	Offy		0	DNS, too slow	0
Art Cross	99	Springfield Welding No.6	Bessie Lee Paoli	Kurtis 4000	Offy		0	DNS, too slow	0
Danny Oakes	99	Micro-Nut No.49	John C. Balch	Kurtis 4000	Offy		0	DNS, too slow	0
Jim Rathmann	99	Elgin Piston Pin No.59	Elgin Piston Pin Co.	Kurtis 500	Offy		0	DNS, too slow	0
Ed Elisian	99	Peter Wales Trucking No.68	Peter Wales	Kurtis 4000	Offy		0	DNS, too slow	0
Duke Nalon	99	Novi No.8	Novi Racing Corp., Inc	Kurtis FD	NoviSc	0		DNS, too slow	0
Bob Scott	99	Ray Brady No.18	Ray T. Brady	Kurtis 4000	Offy		0	Did not qualify	0
Cliff Griffith	99	Tom Sarafoff No.22	Tom Sarafoff	Kurtis 2000	Offy		0	Did not qualify	0
George Tichenor	99	Tom Sarafoff No.22	Tom Sarafoff	Kurtis 2000	Offy		0	Did not qualify	0
Jerry Hoyt	99	Sumar No.23	Chapman S. Root	Kurtis 3000	Offy		0	Did not qualify	0
Al Herman	99	Ed Stone No.36	Edgar S. Stone	Kurtis 3000	Offy		0	Did not qualify	0
George Fonder	99	Ed Stone No.36	Edgar S. Stone	Kurtis 3000	Offy		0	Did not qualify	0
Eddie Russo	99	Federal Engineering No.37	Federal Automotive Assoc.	Kurtis 4000	Offy		0	Did not qualify	0
Bill Holland	99	Bardahl No.38	Ed Walsh	Kurtis 500	Offy		0	Did not qualify	0
Pat Flaherty	99	Shouse Motors No.39	James F. Shouse	Kurtis 500	ChryV8	0		Did not qualify	0
Eddie Sachs	99	Marion Engineering No.54	Milt Marion	Kurtis 2000	Offy		0	Did not qualify	0
Johnny Tolan	99	W. P. T. No.69	Hoosier Racing Team	Kurtis 500	OffySc	0		Did not qualify	0
Pat Flaherty	99	Dunn Engineering No.89	Harry Dunn	Kurtis 500	Offy		0	Did not qualify	0
Billy DeVore	99	Youngstown White Truck No.93	Roy McKay	Kurtis 3000	Offy		0	Did not qualify	0
Lee Wallard	99	Belanger No.99	Murrell Belanger	Kurtis D	Offy		0	Retired	0
Milwaukee, WI — 100 Mile — 6 Jun									
Jack McGrath	4	Hinkle No.2	Jack B. Hinkle	Kurtis 4000	Offy	21	100	Finished	1,917
Gene Hartley	6	John Zink No.31	John S. Zink	Kurtis 4000	Offy	11	100	Finished	1,275
Troy Ruttman	7	Springfield Welding No.6	Bessie Lee Paoli	Kurtis 4000	Offy	4	100	Finished	1,022
Bob Sweikert	11	Lutes Truck Parts No.17	Francis Bardazo	Kurtis 4000	Offy	8	100	Finished	766
Johnny Tolan	12	Anderson No.49	Clarence Anderson	Kurtis 4000	Offy	10	100	Finished	639
Joe Sostilio	13	Bardahl No.45	Ed Walsh	Kurtis 4000	Offy	19	100	Finished	575
Ed Elisian	14	Peter Wales Trucking No.68	Peter Wales	Kurtis 4000	Offy	20	100	Finished	511
Johnnie Parsons	16	Ansted Rotary No.5	Hoosier Racing Team, Inc	Kurtis 500	Offy	17	100	Finished	320
Bob Scott	18	Travelon Trailer No.21	Ernest L. Ruiz	Kurtis 500	Offy	22	100	Finished	255
Duane Carter	19	Auto Shippers No.16	Eugene A. Casaroll	Kurtis 4000	Offy	12	100	Finished	192
Larry Crockett	20	Federal Engineering No.28	Federal Automotive Assoc.	Kurtis 3000	Offy	3	93	Wrecked turn 4	192
Jerry Hoyt	99	Belanger No.99	Murrell Belanger	Kurtis D	Offy		0	DNS, too slow	0
Sam Hanks	99	Bardahl No.1	Ed Walsh	Kurtis 4000	Offy		0	DNS, too slow	0
Tony Bettenhausen	99	Mel Wiggers No.10	Mel B. Wiggers	Kurtis 4000	Offy		0	DNS, too slow	0
Bill Boyd	99	Ray Brady No.18	Ray T. Brady	Kurtis 4000	Offy		0	DNS, too slow	0
Ray Crawford	99	Crawford No.32	Ray Crawford	Kurtis 500	Offy		0	DNS, too slow	0
Marshall Teague	99	Hart Fullerton No.3	Hart Fullerton	Kurtis 4000	Offy		0	Wrecked practice, T3	0
Johnny Fedricks	99	Dunn Engineering No.89	Harry Dunn	Kurtis 500	Offy		0	Wrecked qualifying	0
Langhorne, PA — 100 Mile — 20 Jun									
Sam Hanks	3	Bardahl No.1	Ed Walsh	Kurtis 4000	Offy	13	100	Finished	1,479
Bob Sweikert	4	Lutes Truck Parts No.17	Francis Bardazo	Kurtis 4000	Offy	7	100	Finished	1,076
Joe Sostilio	7	Bardahl No.45	Ed Walsh	Kurtis 4000	Offy	1	66	Flagged	538
Larry Crockett	8	Federal Engineering No.28	Federal Automotive Assoc.	Kurtis 3000	Offy	16	99	Flagged	538

Driver	Fin Pos	Car Name	Entrant	Chassis	Engine	Start Pos	Laps	Time/Speed/Out	Purse
Bob Scott	10	Ray Brady No.18	Ray T. Brady	Kurtis 4000	Offy	9	61	Flagged	403
Ed Elisian	12	Peter Wales Trucking No.68	Peter Wales	Kurtis 4000	Offy	8	75	Spun out	336
Mike Nazaruk	13	McNamara No.73	Lee Elkins	Kurt/Wett	Offy	2	55	Stalled	269
Wally Campbell	98	Ray Brady No.18	Ray T. Brady	Kurtis 4000	Offy	9	33	Rel Scott 62-94	0
Ernie McCoy	98	Bardahl No.45	Ed Walsh	Kurtis 4000	Offy		34	Rel Sostilio 67-100	0
Darlington, SC — 200 Mile — 5 Jul									
Johnny Tolan	4	Anderson No.49	Clarence Anderson	Kurtis 4000	Offy	11	146	1:38:54.77 121.319	975
Jerry Hoyt	5	Belanger No.99	Murrell Belanger	Kurtis D	Offy	6	146	1:39:19.67 120.812	900
Art Cross	6	Springfield Welding No.6	Bessie Lee Paoli	Kurtis 4000	Offy	17	146	1:39:20.66 120.792	787
Joe Sostilio	7	Bardahl No.45	Ed Walsh	Kurtis 4000	Offy	13	146	1:39:33.62 120.530	600
Sam Hanks	8	Bardahl No.1	Ed Walsh	Kurtis 4000	Offy	9	76	1:40:21.43 119.573	487
Duane Carter	9	Auto Shippers No.34	Eugene A. Casaroll	Kurtis 500	Offy	25	145	Flagged	450
Larry Crockett	11	Federal Engineering No.28	Federal Automotive Assoc.	Kurtis 3000	Offy	4	144	Flagged	375
Eddie Russo	17	Peter Wales Trucking No.68	Peter Wales	Kurtis 4000	Offy	24	132	Flagged	150
Bob Scott	20	Ray Brady No.18	Ray T. Brady	Kurtis 4000	Offy	21	108	Wrecked, fatal, BS	75
Jack McGrath	21	Hinkle No.2	Jack B. Hinkle	Kurtis 4000	Offy	5	104	Spun out, turn 3	75
Bob Sweikert	22	Lutes Truck Parts No.17	Francis Bardazo	Kurtis 4000	Offy	16	102	Out	75
Jimmy Daywalt	23	Auto Shippers No.16	Eugene A. Casaroll	Kurtis 4000	Offy	20	59	Wrecked, turn 2	75
Mike Nazaruk	25	McNamara No.73	Lee Elkins	Kurt/WettD	Offy	18	61	Out	75
Jim Rathmann	27	Travelon Trailer No.21	Ernest L. Ruiz	Kurtis 500	Offy	29	46	Lost brakes	75
Johnnie Parsons	30	Ed Stone No.36	Edgar S. Stone	Kurtis 3000	Offy	10	9	Broken crankshaft	75
Len Duncan	98	Bardahl No.1	Ed Walsh	Kurtis 4000	Offy		70	Rel Hanks 77-146	0
Ernie McCoy	98	Auto Shippers No.16	Eugene A. Casaroll	Kurtis 4000	Offy		18	Rel Daywalt 60-78	0
Ray Crawford	99	Crawford No.32	Ray Crawford	Kurtis 500	Offy		0	Did not qualify	0
Springfield, IL — 100 Mile — 21 Aug									
Bob Sweikert	4	Lutes Truck Parts No.17	Francis Bardazo	Kurtis 4000	Offy	2	100	Finished	1,070
Larry Crockett	6	Federal Engineering No.37	Federal Automotive Assoc.	Kurtis 4000	Offy	15	100	Finished	823
Jack McGrath	8	Hinkle No.2	Jack B. Hinkle	Kurtis 4000	Offy	9	100	Finished	617
Mike Nazaruk	9	McNamara No.73	Lee Elkins	Kurt/Wett	Offy	4	100	Finished	576
Johnny Tolan	10	Anderson No.49	Clarence Anderson	Kurtis 4000	Offy	10	100	Finished	535
Ed Elisian	13	Bardahl No.45	Ed Walsh	Kurtis 4000	Offy	3	99	Flagged	411
Danny Oakes	14	Hart Fullerton No.3	Hart Fullerton	Kurtis 4000	Offy	6	99	Flagged	329
Rex Easton	17	Dunn Engineering No.89	Harry Dunn	Kurtis 500	Offy	18	81	Flagged	164
Ray Crawford	18	Crawford No.32	Ray Crawford	Kurtis 4000	Offy	17	93	Flagged	164
Andy Linden	98	Dunn Engineering No.89	Harry Dunn	Kurtis 500	Offy		15	Rel Easton 82-96	0
Gene Hartley	99	John Zink No.31	John S. Zink	Kurtis 4000	Offy		0	DNS, too slow	0
Jerry Hoyt	99	Belanger No.99	Murrell Belanger	Kurtis	Offy		0	DNS, too slow	0
Duane Carter	99	Auto Shippers No.16	Eugene A. Casaroll	Kurtis 4000	Offy		0	DNS, too slow	0
Art Cross	99	Springfield Welding No.6	Bessie Lee Paoli	Kurtis 4000	Offy		0	DNS, too slow	0
Andy Linden	99	Tom Sarafoff No.22	Tom Sarafoff	Kurtis 2000	Offy		0	DNS, too slow	0
Johnnie Parsons	99	Bardahl No.1	Ed Walsh	Kurtis 4000	Offy		0	DNS, too slow	0
Milwaukee, WI — 200 Mile — 29 Aug									
Mike Nazaruk	4	McNamara No.73	Lee Elkins	Kurt/WettD	Offy	22	200	Finished	2,156
Art Cross	5	Springfield Welding No.6	Bessie Lee Paoli	Kurtis 4000	Offy	3	200	Finished	1,771
Gene Hartley	7	John Zink No.31	John S. Zink	Kurtis 4000	Offy	14	199	Flagged	1,232
Larry Crockett	8	Federal Engineering No.37	Federal Automotive Assoc.	Kurtis 4000	Offy	5	199	Flagged	1,155
Fred Agabashian	10	Bardahl No.12	Ed Walsh	Kurtis 4000	Offy	11	198	Flagged	1,001
Ed Elisian	11	Bardahl No.45	Ed Walsh	Kurtis 4000	Offy	16	198	Flagged	924
Pat Flaherty	12	Dunn Engineering No.89	Harry Dunn	Kurtis 500	Offy	10	197	Flagged	770
Duane Carter	13	Auto Shippers No.16	Eugene A. Casaroll	Kurtis 4000	Offy	20	197	Flagged	693
Bill Homeier	15	Sumar No.23	Chapman S. Root	Kurtis 3000	Offy	26	192	Flagged	462
Danny Kladis	18	Tom Sarafoff No.22	Tom Sarafoff	Kurtis 2000	Offy	25	187	Flagged	308
Bob Sweikert	22	Lutes Truck Parts No.17	Francis Bardazo	Kurtis 4000	Offy	12	179	Black flag, too slow	154
Rex Easton	24	Mel Wiggers No.10	Mel B. Wiggers	Kurtis 4000	Offy	9	131	Lost brakes	154
Johnny Tolan	25	Anderson No.49	Clarence Anderson	Kurtis 4000	Offy	24	111	Hub failure	154
Danny Oakes	99	Hart Fullerton No.3	Hart Fullerton	Kurtis 4000	Offy		0	DNS, too slow	0
Jerry Hoyt	99	Belanger No.99	Murrell Belanger	Kurtis D	Offy		0	DNS, too slow	0
Eddie Russo	99	Peter Wales Trucking No.68	Peter Wales	Kurtis 4000	Offy		0	DNS, too slow	0
Curley Boyd	99	Shouse Motors No.39	James F. Shouse	Kurtis 500	ChryV8	0		DNS, too slow	0
Pikes Peak, CO — 12 Mile — 6 Sep									
Keith Andrews	1	Joe Hunt Magneto No.27	Joe Hunt	Kurtis 2000	Offy	18	1	0:14:39.7 50.826	3,500
Larry Crockett	4	Federal Engineering No.37	Federal Automotive Assoc.	Kurtis 4000	Offy	1	1	0:14:58.0 49.791	1,000
Louis Unser	18	Federal Auto Associates No.28		Kurtis	Offy	5	1	0:18:53.8 39.436	0
DuQuoin, IL — 100 Mile — 6 Sep									
Sam Hanks	1	Belanger No.99	Murrell Belanger	Kurtis	Offy	2	83	0:56:34.33 88.029	3,500
Johnnie Parsons	7	Bardahl No.15	Ed Walsh	Kurtis 4000	Offy	7	81	Flagged	560
Bob Sweikert	8	Lutes Truck Parts No.17	Francis Bardazo	Kurtis 4000	Offy	10	81	Flagged	525
Jack McGrath	9	Hinkle No.2	Jack B. Hinkle	Kurtis 4000	Offy	12	81	Flagged	490
Mike Nazaruk	10	McNamara No.73	Lee Elkins	Kurt/Wett	Offy	5	80	Flagged	455
Gene Hartley	12	John Zink No.31	John S. Zink	Kurtis 4000	Offy	11	79	Flagged	350

Driver	Fin Pos	Car Name	Entrant	Chassis	Engine	Start Pos	Laps	Time/Speed/Out	Purse
Danny Kladis	13	Tom Sarafoff No.22	Tom Sarafoff	Kurtis 2000	Offy	14	73	Flagged	315
Ray Crawford	15	Crawford No.32	Ray Crawford	Kurtis 4000	Offy	4	65	Out	245
Duane Carter	18	Auto Shippers No.16	Eugene A. Casaroll	Kurtis 4000	Offy	9	35	Out	140
Rex Easton	99	Bardahl No.45	Ed Walsh	Kurtis 4000	Offy		0	DNS, too slow	0
Eddie Russo	99	Peter Wales Trucking No.68	Peter Wales	Kurtis 4000	Offy		0	DNS, too slow	0
Pat Flaherty	99	Dunn Engineering No.89	Harry Dunn	Kurtis 500	Offy		0	DNS, too slow	0
Danny Oakes	99	Hart Fullerton No.3	Hart Fullerton	Kurtis 4000	Offy		0	DNS, too slow	0
Duke Nalon	99	Mel Wiggers No.10	Mel B. Wiggers	Kurtis 4000	Offy		0	DNS, too slow	0
Johnny Tolan	99	Anderson No.49	Clarence Anderson	Kurtis 4000	Offy		0	DNS, too slow	0
Bill Homeier	99	Sumar No.23	Chapman S. Root	Kurtis 3000	Offy		0	DNS, too slow	0
Curley Boyd	99	Shouse Motors No.39	James F. Shouse	Kurtis 500	ChryV8	0		DNS, too slow	0
Syracuse, NY — 100 Mile — 11 Sep									
Bob Sweikert	1	Lutes Truck Parts No.17	Francis Bardazo	Kurtis 4000	Offy	3	100	1:06:25 90.320	3,750
Johnny Tolan	3	Anderson No.49	Clarence Anderson	Kurtis 4000	Offy	5	100	Finished	1,500
Ed Elisian	5	Schaefer Gear Works No.68		Kurtis 4000	Offy	14	100	Finished	900
Andy Linden	7	Bardahl No.45	Ed Walsh	Kurtis 4000	Offy	2	100	Finished	600
Johnnie Parsons	8	Bardahl No.15	Ed Walsh	Kurtis 4000	Offy	8	99	Flagged	562
Jack McGrath	10	Hinkle No.2	Jack B. Hinkle	Kurtis 4000	Offy	7	99	Flagged	487
Jimmy Davies	11	Auto Shippers No.16	Eugene A. Casaroll	Kurtis 4000	Offy	15	98	Flagged	450
Larry Crockett	13	Federal Engineering No.37	Federal Automotive Assoc.	Kurtis 4000	Offy	17	98	Flagged	337
Al Keller	14	Springfield Welding No.6	Bessie Lee Paoli	Kurtis 4000	Offy	9	97	Flagged	300
Mike Nazaruk	17	McNamara No.73	Lee Elkins	Kurt/Wett	Offy	18	40	Brake failure	150
Jimmy Davies	99	Crawford No.32	Ray Crawford	Kurtis 4000	Offy		0	DNS, too slow	0
Indianapolis, IN — 100 Mile — 18 Sep									
Bob Sweikert	2	Lutes Truck Parts No.17	Francis Bardazo	Kurtis 4000	Offy	2	100	Finished	4,248
Sam Hanks	3	Belanger No.99	Murrell Belanger	Kurtis	Offy	4	100	Finished	2,360
Jack McGrath	4	Hinkle No.2	Jack B. Hinkle	Kurtis 4000	Offy	6	100	Finished	1,534
Ed Elisian	6	Springfield Welding No.6	Bessie Lee Paoli	Kurtis 4000	Offy	7	99	Flagged	1,180
Johnny Tolan	8	Anderson No.49	Clarence Anderson	Kurtis 4000	Offy	12	98	Flagged	885
Jerry Hoyt	10	Bardahl No.45	Ed Walsh	Kurtis 4000	Offy	10	97	Flagged	708
Rex Easton	12	Mel Wiggers No.10	Mel B. Wiggers	Kurtis 4000	Offy	14	95	Flagged	590
Ray Crawford	14	Crawford No.32	Ray Crawford	Kurtis 4000	Offy	17	92	Flagged	472
Johnnie Parsons	16	Walsh No.15	Ed Walsh	Kurtis 4000	Offy	16	72	Spun out, turn 1	354
Danny Kladis	99	Tom Sarafoff No.22	Tom Sarafoff	Kurtis 2000	Offy		0	DNS, too slow	0
Danny Oakes	99	Hart Fullerton No.3	Hart Fullerton	Kurtis 4000	Offy		0	DNS, too slow	0
Gene Hartley	99	John Zink No.31	John S. Zink	Kurtis 4000	Offy		0	DNS, too slow	0
Larry Crockett	99	Federal Engineering No.37	Federal Automotive Associates	Kurtis 4000	Offy		0	DNS, too slow	0
Duane Carter	99	Auto Shippers No.16	Eugene A. Casaroll	Kurtis 4000	Offy		0	DNS, too slow	0
Pat Flaherty	99	Dunn Engineering No.89	Harry Dunn	Kurtis 500	Offy		0	DNS, too slow	0
Curley Boyd	99	Shouse Motors No.95	James F. Shouse	Kurtis 500	ChryV8	0		DNS, too slow	0
Bill Homeier	99	Sumar No.23	Chapman S. Root	Kurtis 3000	Offy		0	Did not qualify	0
Elmer George	99	Peter Wales Trucking No.68	Peter Wales	Kurtis 4000	Offy		0	Did not qualify	0
Sacramento, CA — 100 Mile — 17 Oct									
Jack McGrath	2	Hinkle No.2	Jack B. Hinkle	Kurtis 4000	Offy	1	100	Finished	1,839
Bob Sweikert	3	Lutes Truck Parts No.17	Francis Bardazo	Kurtis 4000	Offy	7	100	Finished	1,021
Johnny Tolan	9	Anderson No.49	Clarence Anderson	Kurtis 4000	Offy	11	95	Flagged	357
Keith Andrews	10	Joe Hunt Magneto No.27	Joe Hunt	Kurtis 2000	Offy	18	93	Flagged	332
Rodger Ward	13	M. A. Walker No.82	M.A. Walker	Kurtis	Offy	10	44	Broken rear end	229
Sam Hanks	15	Belanger No.1	Murrell Belanger	Kurtis	Offy	5	32	Hit by rock	178
Jerry Hoyt	16	Bardahl No.45	Ed Walsh	Kurtis 4000	Offy	16	21	Blew engine	153
Eddie Sachs	99	John Zink No.31	John S. Zink	Kurtis 4000	Offy		0	DNS, too slow	0
Danny Oakes	99	Hart Fullerton No.3	Hart Fullerton	Kurtis 4000	Offy		0	DNS, too slow	0
Don Olds	99	Dobry No.63	Leo Dobry	Kurtis 2000	Offy		0	DNS, too slow	0
Ray Crawford	99	Crawford No.32	Ray Crawford	Kurtis 4000	Offy		0	DNS, too slow	0
Phoenix, AZ — 100 Mile — 7 Nov									
Bob Sweikert	3	Lutes Truck Parts No.17	Francis Bardazo	Kurtis 4000	Offy	2	99	Flagged	1,000
Sam Hanks	5	Belanger No.1	Murrell Belanger	Kurtis	Offy	3	99	Flagged	600
Eddie Sachs	7	John Zink No.31	John S. Zink	Kurtis 4000	Offy	13	96	Flagged	400
Shorty Templeman	8	Schaefer Gear Works No.68		Kurtis 4000	Offy	1	96	Flagged	375
Jerry Hoyt	11	Bardahl No.45	Ed Walsh	Kurtis 4000	Offy	14	94	Flagged	300
Tom Hinnershitz	13	Springfield Welding No.6	Bessie Lee Paoli	Kurtis 4000	Offy	6	91	Flagged	225
Ray Crawford	14	Crawford No.32	Ray Crawford	Kurtis 4000	Offy	15	89	Flagged	200
Larry Crockett	15	Federal Engineering No.37	Federal Automotive Assoc.	Kurtis 4000	Offy	17	73	Flagged	175
Johnny Tolan	18	Anderson No.49	Clarence Anderson	Kurtis 4000	Offy	11	36	Burned wheel bearing	100
Jack McGrath	99	Hinkle No.2	Jack B. Hinkle	Kurtis 4000	Offy		0	DNS, too slow	0
Bill Homeier	99	Hart Fullerton No.3	Hart Fullerton	Kurtis 4000	Offy		0	DNS, too slow	0
Rodger Ward	99	M. A. Walker No.35	M.A. Walker	Kurtis	Offy		0	DNS, too slow	0
Duane Carter	99	McNamara No.73	Lee Elkins	Kurtis 2000	Offy		0	DNS, too slow	0
Johnnie Parsons	99	Bardahl No.15	Ed Walsh	Kurtis 4000	Offy		0	DNS, too slow	0
Duke Nalon	99	Dobry No.63	Leo Dobry	Kurtis 2000	Offy		0	DNS, too slow	0

Driver	Fin Pos	Car Name	Entrant	Chassis	Engine	Start Pos	Laps	Time/Speed/Out	Purse
Las Vegas, NV — 100 Mile — 14 Nov									
Rodger Ward	4	Grant Piston Ring No.35	M.A. Walker	Kurtis	Offy	16	99	Finished	650
Tom Hinnershitz	6	Crawford No.32	Ray Crawford	Kurtis 4000	Offy	10	98	Flagged	500
Rex Easton	7	Springfield Welding No.6	Bessie Lee Paoli	Kurtis 4000	Offy	8	98	Flagged	400
Larry Crockett	10	Federal Engineering No.37	Federal Automotive Assoc.	Kurtis 4000	Offy	3	93	Wrecked, turn 4	325
Johnny Tolan	12	Anderson No.49	Clarence Anderson	Kurtis 4000	Offy	17	86	Wrecked, turn 4	250
Bob Sweikert	14	Lutes Truck Parts No.17	Francis Bardazo	Kurtis 4000	Offy	4	2	Wrecked	200
Sam Hanks	99	Belanger No.1	Murrell Belanger	Kurtis	Offy		0	DNS, too slow	0
Duane Carter	99	McNamara No.73	Lee Elkins	Kurt/Wett	Offy		0	DNS, too slow	0
Johnnie Parsons	99	Bardahl No.15	Ed Walsh	Kurtis 4000	Offy		0	Wrecked practice	0
Bill Homeier	99	Hart Fullerton No.3	Hart Fullerton	Kurtis 4000	Offy		0	DNS, too slow	0
Gene Hartley	99	Commercial Motor Freight No.62	Karl Hall	Kurtis 4000	Offy		0	DNS, too slow	0
Don Freeland	99	John Zink No.31	John S. Zink	Kurtis 4000	Offy		0	DNS, too slow	0
Shorty Templeman	99	Schaefer Gear Works No.68		Kurtis 4000	Offy		0	DNS, too slow	0
Jerry Hoyt	99	Bardahl No.45	Ed Walsh	Kurtis 4000	Offy		0	DNS, too slow	0
Duke Nalon	99	Dobry No.63	Leo Dobry	Kurtis 2000	Offy		0	Did not qualify	0
Williams Grove, PA — 25 Mile — 25 Jul									
Johnny Tolan	7	Anderson No.49	Clarence Anderson	Kurtis 4000	Offy	4	50	Finished	450
Bob Sweikert	10	Lutes Truck Parts No.17	Francis Bardazo	Kurtis 4000	Offy	2	50	Finished	337
Johnnie Parsons	13	Bardahl No.45	Ed Walsh	Kurtis 4000	Offy	9	42	Flagged	225
Eddie Russo	99	Peter Wales Trucking No.68	Peter Wales	Kurtis 4000	Offy	20	0	Did not start	0
1955									
Indianapolis, IN — 500 Mile — 30 May									
Bob Sweikert	1	John Zink No.6	John Zink Co.	Kurtis 500D	Offy	14	200	3:53:59.53 128.209	76,138
Tony Bettenhausen	2	Chapman No.10	H. A. Chapman	Kurtis 500C	Offy	2	123	3:56:43.11 126.733	30,088
Jimmy Davies	3	Bardahl\Clancy No.15	Pat Clancy	Kurtis 500B	Offy	10	200	3:57:31.89 126.299	16,988
Walt Faulkner	5	Merz Engineering No.77	Merz Engineering, Inc	Kurtis 500C	Offy	7	176	3:59:16.66 125.377	10,763
Andy Linden	6	Massaglia Hotel No.19	Joseph Massaglia, Jr	Kurtis 4000D	Offy	8	200	3:59:57.47 125.022	8,513
Pat O'Connor	8	Ansted Rotary No.29	Rotary Engineering Corp.	Kurtis 500D	Offy	19	200	4:00:41.09 124.644	6,413
Jimmy Daywalt	9	Sumar No.48	Chapman S. Root	Kurtis R	Offy	17	200	4:01:09.39 124.401	6,413
Pat Flaherty	10	Dunn Engineering No.89	Harry Dunn	Kurtis 500B	Offy	12	200	4:01:46.05 124.086	6,113
Chuck Weyant	12	Federal Engineering No.41	Federal Automotive Assoc.	Kurtis 3000D	Offy	25	196	Flagged	3,613
Cal Niday	16	D-A Lubricants No.22	Racing Associates	Kurtis 500B	Offy	9	170	Wrecked NW	3,483
Art Cross	17	Belanger No.99	Murrell Belanger	Kurtis 500D	Offy	24	168	Spun out	6,663
Sam Hanks	19	Jones & Maley No.8	Cars, Inc	Kurtis 500C	Offy	6	134	Transmission	2,913
Johnnie Parsons	21	Trio Brass No.16	Carl Lee Gehlhausen	Kurtis 500D	Offy	27	119	Magneto	2,653
Ray Crawford	23	Crawford No.49	Ray Crawford	Kurtis 500B	Offy	23	111	Engine stalled	2,893
Bill Vukovich	25	Hopkins No.4	Motor Racers Inc	Kurtis 500C	Offy	5	56	Wrecked BS - fatal	10,833
			Lindsey Hopkins						
Jack McGrath	26	Hinkle No.3	Jack B. Hinkle	Kurtis 500C	Offy	3	54	Broken oil line	6,353
Al Keller	27	Sam Traylor No.42	Samuel W. Traylor, III	Kurtis 2000D	Offy	22	54	Wrecked BS	2,873
Johnny Boyd	29	Sumar No.39	Chapman S. Root	Kurtis 500C	Offy	26	53	Wrecked BS	2,413
Ed Elisian	30	Peter Wales\Westwood Tool No.68	Peter Wales	Kurtis 4000D	Offy	29	53	Stopped for wreck	2,733
Fred Agabashian	32	Federal Engineering No.14	Federal Automotive Assoc.	Kurtis 500D	Offy	4	39	Spun out, NE turn	2,723
Paul Russo	98	Chapman No.10	H. A. Chapman	Kurtis 500C	Offy		77	Rel Bettenhsn 57-133	0
Bill Homeier	98	Merz Engineering No.77	Merz Engineering, Inc	Kurtis 500C	Offy		24	Rel Faulkner 138-161	0
Len Duncan	99	Ray Brady No.24	Ray T. Brady	Kurtis 4000D	Offy		0	DNS, too slow	0
Spider Webb	99	Ed Walsh No.9	Ed Walsh	Kurtis 500C	Offy		0	Did not qualify	0
Johnnie Parsons	99	Ed Walsh No.9	Ed Walsh	Kurtis 500C	Offy		0	Did not qualify	0
Troy Ruttman	99	Novi Air Conditioner No.18	Novi Racing Corp.	Kurtis FD	NoviSc	0		Did not qualify	0
Paul Russo	99	Wolcott No.21	Roger G. Wolcott	Kurtis 500A	OffySc	0		Wrecked practice	0
Ed Elisian	99	Lutes Truck Parts No.25	Francis Bardazon	Kurtis 4000D	Offy		0	Did not qualify	0
Tony Bonadies	99	Donaldson No.36	Duke Donaldson	Kurtis 3000D	Offy		0	Did not qualify	0
LeRoy Warriner	99	Ansted Rotary No.64	Rotary Engineering Corp.	Kurtis 500A	Offy		0	Did not qualify	0
Bill Homeier	99	Travelon Trailer No.72	Ernest L. Ruiz	Kurtis 500B	Offy		0	Did not qualify	0
Len Duncan	99	McNamara No.73	Kalamazoo Sports	Kurtis 500C	Offy		0	Wrecked practice	0
			Lee Elkins						
Manuel Ayulo	99	Peter Schmidt No.88	Peter Schmidt	Kurtis 500C	Offy		0	Wrecked - fatal	0
Danny Kladis	99	Roy McKay No.93	Roy McKay	Kurtis 3000D	Offy		0	Did not qualify	0
Milwaukee, WI — 100 Mile — 5 Jun									
Pat Flaherty	3	Dunn Engineering No.89	Harry Dunn	Kurtis 500B	Offy	16	100	Finished	2,160
Andy Linden	6	Massaglia Hotel No.19	Joseph Massaglia, Jr	Kurtis 4000D	Offy	8	100	Finished	1,080
Jack McGrath	8	Hinkle No.3	Jack B. Hinkle	Kurtis 4000D	Offy	12	100	Finished	810
Jimmy Reece	9	Ed Walsh No.9	Ed Walsh	Kurtis 4000D	Offy	14	100	Finished	756
Shorty Templeman	11	Peter Wales\Westwood Tool No.68	Peter Wales	Kurtis 4000D	Offy	9	100	Finished	648
Mike Magill	13	Unverdorben No.95		Kurtis 4000D	Offy	19	100	Finished	486
Rex Easton	17	Mel Wiggers No.47	Mel Wiggers	Kurtis 4000D	Offy	22	97	Flagged	316
Walt Faulkner	21	McNamara No.73	Kalamazoo Sports	Kurt/WettD	Offy	17	37	Brkn left rear hub	108
			Lee Elkins						
Ed Elisian	99	Lutes Truck Parts No.25	Francis Bardazon	Kurtis 4000D	Offy		0	DNS, too slow	20
Bill Homeier	99	Sumar No.51	Chapman S. Root	Kurtis 3000D	Offy		0	DNS, too slow	20
Johnny Tolan	99	Trio Brass No.16	Carl Lee Gehlhausen	Kurtis 4000D	Offy		0	DNS, too slow	60

Kurtis-Kraft

Driver	Fin Pos	Car Name	Entrant	Chassis	Engine	Start Pos	Laps	Time/Speed/Out	Purse
Pat O'Connor	18	Ansted Rotary No.7	Ansted Rotary Corp.	Kurtis 500D	Offy	3	187	Magneto	8,924
Jim Rathmann	20	Hopkins No.24	Lindsey Hopkins	Kurtis 500C	Offy	2	175	Oil trouble	3,564
Johnny Tolan	21	Trio Brass No.34	Carl L. Anderson	Kurtis 500D	Offy	31	173	Flagged	3,084
Tony Bettenhausen	22	Belanger No.99	Murrell Belanger	Kurtis 500D	Offy	5	160	Wrecked SW	2,754
Ed Elisian	23	Hoyt Machine No.10	Fred Somers	Kurtis 500C	Offy	14	123	Lost brakes	2,624
Jimmy Daywalt	24	Sumar No.48	Chapman S. Root	Kurtis 500C	Offy	16	134	Wrecked SE	2,564
Jack Turner	25	Travelon Trailer No.54	Ernest L. Ruiz	Kurtis 500B	Offy	24	131	Engine	2,564
Keith Andrews	26	Dunn Engineering No.89	Harry Dunn	Kurtis 500B	Offy	20	94	Broken transmission	2,834
Andy Linden	27	Chapman No.5	H. A. Chapman	Kurtis 500C	Offy	9	90	Oil leak	2,504
Al Herman	28	Bardahl/Clancy No.12	Pat Clancy	Kurtis 500B	Offy	27	74	Wrecked MS	2,474
Ray Crawford	29	Crawford No.49	Ray Crawford	Kurtis 500B	Offy	17	49	Wrecked NW	2,444
Johnny Boyd	30	Bowes Seal Fast No.15	George Bignotti	Kurtis 500G	Offy	12	35	Oil leak	2,414
Troy Ruttman	31	John Zink No.53	John S. Zink	Kurtis 500D	Offy	11	22	Spun out SW	2,384
Paul Russo	33	Novi Vespa No.29	Novi Racing Corp., Inc	Kurtis 500F	NoviSc	8	21	Wrecked SW	3,974
Eddie Russo	98	Hoyt Machine No.10	Fred Somers	Kurtis 500C	Offy		37	Rel Elisian 124-160	0
Nino Farina	99	Bardahl-Ferrari No.9	Bardahl Lubricants	Kurtis 500D	Ferrar	0		Did not qualify	0
Tony Bonadies	99	Donaldson No.25	Duke Donaldson	Kurtis 3000D	Offy		0	Did not qualify	0
Jimmy Davies	99	Novi Air Conditioning No.31	Novi Racing Corp., Inc	Kurtis 500F	NoviSc	0		Did not qualify	0
Johnny Kay	99	Peter Wales No.35	Peter Wales	Kurtis 4000D	Offy		0	Did not qualify	0
Marshall Teague	99	Sumar No.47	Chapman S. Root	Kurtis	Offy		0	Did not qualify	0
Len Duncan	99	Ray Brady No.51	Ray T. Brady	Kurtis 4000D	Offy		0	Did not qualify	0
Gig Stephens	99	Slick Airway No.85	Otto & Adolph Stoye	Kurtis 2000D	Offy		0	Did not qualify	0
Dick Reese	99	McDonald No.87	Cecil McDonald	Kurtis 4000D	Offy		0	Did not qualify	0
Bill Cheesbourg	99	McDonald No.87	Cecil McDonald	Kurtis 4000D	Offy		0	Did not qualify	0
LeRoy Warriner	99	McKay's Bulldog No.93	Roy McKay	Kurtis 3000D	Offy		0	Did not qualify	0
Milwaukee, WI — 100 Mile — 10 Jun									
Johnny Boyd	5	Bowes Seal Fast No.15	George Bignotti	Kurtis 500G	Offy	1	100	Finished	1,317
Jimmy Reece	7	Massaglia Hotel No.26	Joseph Massaglia, Jr	Kurtis 4000D	Offy	19	100	Finished	878
Al Keller	9	Traylor No.55	Samuel W. Traylor, III	Kurtis 4000D	Offy	21	100	Finished	868
Bob Veith	10	Federal Engineering No.14	Federal Automotive Assoc.	Kurtis 4000D	Offy	5	100	Finished	713
Keith Andrews	12	Dunn Engineering No.89	Harry Dunn	Kurtis 500B	Offy	15	98	Flagged	549
Ed Elisian	99	Lutes Truck Parts No.25	W.J. Lutes	Kurtis 4000D	Offy		0	DNS, too slow	20
Johnny Kay	99	Peter Wales No.35	Peter Wales	Kurtis 4000D	Offy		0	DNS, too slow	35
Andy Linden	99	Shannon Bros. No.64	R.B. & E.E. Shannon	Kurtis 500A	Offy		0	DNS, too slow	60
Johnny Tolan	99	Trio Brass No.34	Carl L. Anderson	Kurtis 4000D	Offy		0	DNS, too slow	0
Langhorne, PA — 100 Mile — 24 Jun									
Al Keller	3	Traylor No.55	Samuel W. Traylor, III	Kurtis 4000	Offy	10	100	Finished	1,850
Ed Elisian	4	Lutes Truck Parts No.25	W.J. Lutes	Kurtis 4000	Offy	15	100	Finished	1,203
Mike Magill	5	Unverdorben No.24		Kurtis 4000	Offy	12	100	Finished	1,110
Jimmy Reece	9	Hart Fullerton No.26	Hart Fullerton	Kurtis 4000	Offy	3	97	Flagged	648
Bob Veith	10	Federal Engineering No.14	Federal Automotive Associates	Kurtis 4000	Offy	2	36	Flagged	601
Rodger Ward	98	Federal Engineering No.14	Federal Automotive Assoc.	Kurtis 4000	Offy		58	Rel Veith 37-94	0
Rodger Ward	99	Holynski No.84		Kurtis	Offy		0	DNS, too slow	0
Darlington, SC — 200 Mile — 4 Jul									
Johnny Boyd	3	Bowes Seal Fast No.15	George Bignotti	Kurtis 500G	Offy	2	146	Finished	1,500
Jimmy Reece	12	Hart Fullerton No.25	Hart Fullerton	Kurtis 4000D	Offy	7	143	Flagged	375
Gene Force	20	Peter Wales No.35	Peter Wales	Kurtis 4000D	Offy	10	133	Flagged	112
Eddie Sachs	21	Filter Queen No.19	Ed Walsh	Kurtis 500C	Offy	11	128	Spun out, turn 4	75
Bob Veith	99	Federal Engineering No.14	Federal Automotive Assoc.	Kurtis 4000D	Offy		0	Wrecked practice, T2	0
Al Keller	99	Traylor No.55	Samuel W. Traylor, III	Kurtis 4000D	Offy		0	Did not qualify	0
Duke Dinsmore	99	Shannon Bros. No.64	R.B. & E.E. Shannon	Kurtis 500A	Offy		0	Wrecked practice	0
Atlanta, GA — 100 Mile — 14 Jul									
Al Keller	2	Traylor No.55	Samuel W. Traylor, III	Kurtis 4000	Offy	17	100	Finished	1,350
Jimmy Reece	7	Massaglia Hotel No.26	Joseph Massaglia, Jr	Kurtis 4000	Offy	16	98	Flagged	300
Bob Carpenter	12	Peter Wales No.35	Peter Wales	Kurtis 4000	Offy	11	93	Flagged	187
Ed Elisian	18	Lutes Truck Parts No.25	W.J. Lutes	Kurtis 4000	Offy	4	49	Magneto	75
Springfield, IL — 100 Mile — 19 Aug									
Eddie Sachs	3	Ray Brady No.51	Ray T. Brady	Kurtis 4000	Offy	6	100	Finished	1,610
Al Keller	4	Traylor No.55	Samuel W. Traylor, III	Kurtis 4000	Offy	9	100	Finished	1,047
Johnny Boyd	7	Bowes Seal Fast No.15	George Bignotti	Kurtis 500G	Offy	5	100	Finished	644
Jim McWithey	9	Dayton Steel No.96	George Walther, Jr	Kurtis 4000	Offy	17	98	Flagged	563
Roy Graham	14	Holynski No.94		Kurtis	Offy	4	97	Flagged	322
Bob Veith	15	Federal Engineering No.14	Federal Automotive Assoc.	Kurtis 4000	Offy	14	97	Flagged	281
Ed Elisian	99	Lutes Truck Parts No.25	W.J. Lutes	Kurtis 4000	Offy		0	DNS, too slow	0
Earl Motter	99	Hart Fullerton No.26	Hart Fullerton	Kurtis 4000			0	DNS, too slow	0
Milwaukee, WI — 250 Mile — 26 Aug									
Jimmy Reece	2	Dunn Engineering No.89	Harry Dunn	Kurtis 500B	Offy	13	250	Finished	5,258

Driver	Fin Pos	Car Name	Entrant	Chassis	Engine	Start Pos	Laps	Time/Speed/Out	Purse
Eddie Sachs	4	Ray Brady No.51	Ray T. Brady	Kurtis 4000D	Offy	24	250	Finished	2,537
Johnny Boyd	5	Bowes Seal Fast No.15	George Bignotti	Kurtis 500G	Offy	1	250	Finished	1,979
Bob Veith	8	Federal Engineering No.14	Federal Automotive Assoc.	Kurtis 4000D	Offy	3	249	Flagged	1,279
Earl Motter	9	Hart Fullerton No.26	Hart Fullerton	Kurtis 4000D	Offy	5	249	Flagged	1,179
Al Keller	12	Traylor No.55	Samuel W. Traylor, III	Kurtis 4000D	Offy	25	247	Flagged	781
Gene Hartley	16	Mel Wiggers No.47	Mel Wiggers	Kurtis 4000D	Offy	6	244	Flagged	398
Duke Dinsmore	18	Shannon Bros. No.64	R.B. & E.E. Shannon	Kurtis 500A	Offy	21	233	Flagged	323
Shorty Templeman	19	Lutes Truck Parts No.25	W.J. Lutes	Kurtis 4000D	Offy	22	210	Magneto	247
Jim McWithey	99	Dayton Steel No.96	George Walther, Jr	Kurtis 4000D	Offy		0	Did not qualify	0
DuQuoin, IL — 100 Mile — 3 Sep									
Bob Veith	4	Federal Engineering No.14	Federal Automotive Assoc.	Kurtis 4000	Offy	16	100	Finished	951
Johnny Boyd	11	Bowes Seal Fast No.15	George Bignotti	Kurtis 500G	Offy	17	99	Flagged	439
Earl Motter	14	Hart Fullerton No.26	Hart Fullerton	Kurtis 4000	Offy	18	99	Flagged	293
Charles Musselman	16	Unverdorben No.95		Kurtis 4000	Offy	12	96	Flagged	219
Gene Hartley	99	Mel Wiggers No.47	Mel Wiggers	Kurtis 4000	Offy		0	DNS, too slow	0
Eddie Sachs	99	Ray Brady No.51	Ray T. Brady	Kurtis 4000	Offy		0	DNS, too slow	0
Shorty Templeman	99	Lutes Truck Parts No.25	W.J. Lutes	Kurtis 4000	Offy		0	DNS, too slow	0
Johnny Kay	99	Peter Wales No.35	Peter Wales	Kurtis 4000	Offy		0	DNS, too slow	0
Jim McWithey	99	Dayton Steel Foundry No.96	George Walther, Jr	Kurtis 4000	Offy		0	DNS, too slow	0
Johnnie Parsons	99	Traylor No.55	Samuel W. Traylor, III	Kurtis 4000	Offy		0	DNS, too slow	0
Syracuse, NY — 100 Mile — 8 Sep									
Johnny Boyd	4	Bowes Seal Fast No.15	George Bignotti	Kurtis 500G	Offy	4	99	Flagged	975
Earl Motter	7	Hart Fullerton No.26	Hart Fullerton	Kurtis 4000	Offy	1	99	Flagged	600
Eddie Sachs	12	Ray Brady No.51	Ray T. Brady	Kurtis 4000	Offy	14	98	Flagged	375
Bob Veith	17	Federal Engineering No.14	Federal Automotive Assoc.	Kurtis 4000	Offy	12	52	Wrecked turn 3	150
Charles Musselman	99	Traylor No.55	Samuel W. Traylor, III	Kurtis 4000	Offy		0	DNS, too slow	0
Van Johnson	99	Peter Wales No.35	Peter Wales	Kurtis 4000	Offy		0	DNS, too slow	0
Speed McFee	99	Unverdorben No.95		Kurtis 4000	Offy		0	DNS, too slow	0
Roy Graham	99	Holynski No.84		Kurtis	Offy		0	DNS, too slow	0
Indianapolis, IN — 100 Mile — 15 Sep									
Gene Hartley	6	Mel Wiggers No.47	Mel Wiggers	Kurtis 4000	Offy	4	100	Finished	1,270
Bob Veith	7	Federal Engineering No.14	Federal Automotive Assoc.	Kurtis 4000	Offy	18	99	Flagged	1,016
Mike Magill	10	Dayton Steel No.96	George Walther, Jr	Kurtis 4000	Offy	9	99	Flagged	825
Johnny Boyd	11	Bowes Seal Fast No.15	George Bignotti	Kurtis 500G	Offy	17	99	Flagged	762
Earl Motter	13	Hart Fullerton No.26	Hart Fullerton	Kurtis 4000	Offy	14	94	Spun out, turn 1	571
Eddie Sachs	99	Ray Brady No.51	Ray T. Brady	Kurtis 4000	Offy		0	DNS, too slow	0
Ed Elisian	99	Lutes Truck Parts No.25	W.J. Lutes	Kurtis 4000	Offy		0	DNS, too slow	0
Ray Crawford	99	Crawford No.49	Ray Crawford	Kurtis 4000	Offy		0	DNS, too slow	0
Bill Cheesbourg	99	Peter Wales No.35	Peter Wales	Kurtis 4000	Offy		0	DNS, too slow	0
Sacramento, CA — 100 Mile — 21 Oct									
Bob Veith	4	Federal Engineering No.14	Federal Automotive Assoc.	Kurtis 4000	Offy	6	100	Finished	765
Johnny Boyd	8	Bowes Seal Fast No.15	George Bignotti	Kurtis 500G	Offy	9	98	Flagged	441
Andy Linden	11	Hart Fullerton No.26	Hart Fullerton	Kurtis 4000	Offy	10	96	Flagged	353
Tony Bettenhausen	14	Dayton Steel No.96	George Walther, Jr	Kurtis 4000	Offy	15	61	Spun out, turn 1	235
Art Bisch	17	Whittington No.87		Kurtis 4000	Offy	8	16	Wrecked, turn 1	117
Mel McGaughy	99	Leo Dobry No.65	Leo Dobry	Kurtis 2000	Offy		0	DNS, too slow	0
Dempsey Wilson	99	Joe Hunt Magneto No.27	Joe Hunt	Kurtis 2000	Offy		0	DNS, too slow	0
Ray Crawford	99	Crawford No.49	Ray Crawford	Kurtis 4000	Offy		0	DNS, too slow	0
Phoenix, AZ — 100 Mile — 12 Nov									
Andy Linden	4	Hart Fullerton No.26	Hart Fullerton	Kurtis 4000	Offy	7	100	Finished	726
Johnny Boyd	6	Bowes Seal Fast No.15	George Bignotti	Kurtis 500G	Offy	16	100	Finished	559
Bob Veith	8	Federal Engineering No.14	Federal Automotive Assoc.	Kurtis 4000	Offy	12	100	Finished	419
Tony Bettenhausen	9	Dayton Steel No.96	George Walther, Jr	Kurtis 4000	Offy	11	100	Finished	391
Art Bisch	10	Whittington No.87		Kurtis 4000	Offy	8	100	Finished	363
Dempsey Wilson	12	Joe Hunt Magneto No.27	Joe Hunt	Kurtis 2000	Offy	13	95	Flagged	279
Mel McGaughy	15	Leo Dobry No.63	Leo Dobry	Kurtis 2000	Offy	15	73	Out	195
Eddie Sachs	99	Traylor No.55	Samuel W. Traylor, III	Kurtis 4000	Offy		0	DNS, too slow	0
Ray Crawford	99	Crawford No.49	Ray Crawford	Kurtis 4000	Offy		0	DNS, too slow	0
Williams Grove, PA — 25 Mile — 22 Jul									
Ed Elisian	3	Lutes Truck Parts No.25	W.J. Lutes	Kurtis 4000	Offy	1	0	Finished	787
Jimmy Davies	5	Unverdorben No.95		Kurtis 4000	Offy	7	0	Finished	418
Jimmy Reece	6	Ray Brady No.51	Ray T. Brady	Kurtis 4000	Offy	2	0	Finished	325
Rodger Ward	7	Holynski No.94		Kurtis	Offy	3	0	Finished	450
Gays Biro	13	Peter Wales No.35	Peter Wales	Kurtis 4000	Offy	12	0		300
Al Keller	99	Traylor No.55	Samuel W. Traylor, III	Kurtis 4000	Offy	20	0	Did not start	0
Earl Motter	99	Hart Fullerton No.26	Hart Fullerton	Kurtis 4000	Offy	22	0	Did not start	0
Dayton, OH — 25 Mile — 12 Aug									
Eddie Sachs	4	Ray Brady No.51	Ray T. Brady	Kurtis 4000	Offy	6	50	Finished	650

Driver	Fin Pos	Car Name	Entrant	Chassis	Engine	Start Pos	Laps	Time/Speed/Out	Purse
Earl Motter	5	Hart Fullerton No.26	Hart Fullerton	Kurtis 4000	Offy	12	50	Finished	525
Gene Force	9	Peter Wales No.35	Peter Wales	Kurtis 4000	Offy	10	50	Finished	700
Jim McWithey	14	Dayton Steel Foundry No.96	George Walther, Jr	Kurtis 4000	Offy	4	50	Finished	300
1957									
Indianapolis, IN — 500 Mile — 30 May									
Paul Russo	4	Novi Air Conditioning No.54	Novi Racing Cars, Inc	Kurtis 500F	NoviSc	10	200	3:44:11.10 133.818	19,369
Andy Linden	5	McNamara No.73	Kalamazoo Sports, Inc Lee Elkins	Kurtis 500G	Offy	12	200	3:44:28.55 133.645	11,094
Johnny Boyd	6	Bowes Seal Fast No.6	George Bignotti	Kurtis 500G	Offy	5	200	3:45:49.55 132.846	8,194
Marshall Teague	7	Sumar No.48	Chapman S. Root	Kurtis 500C	Offy	28	200	3:45:59.85 132.745	6,819
Pat O'Connor	8	Sumar No.12	Chapman S. Root	Kurtis 500G	Offy	1	200	3:46:47.35 132.281	8,619
Jack Turner	11	Bardahl No.19	Pat Clancy	Kurtis 500G	Offy	19	200	3:49:10.30 130.906	3,639
Bob Christie	13	Jones & Maley No.95	Cars, Inc	Kurtis 500C	Offy	33	197	Flagged 129.167	3,299
Chuck Weyant	14	Central Excavating No.82	Pete Salemi	Kurtis 500C	Offy	25	196	Flagged 128.078	3,429
Tony Bettenhausen	15	Novi Air Conditioning No.27	Novi Racing Cars, Inc	Kurtis 500F	NoviSc	22	195	Flagged 126.735	4,059
Johnnie Parsons	16	Sumar No.18	Chapman S. Root	Kurtis 500G	Offy	17	195	Flagged 126.274	2,989
Don Freeland	17	Ansted Rotary No.3	Ansted Rotary, Inc	Kurtis 500D	Offy	21	192	Flagged 125.763	2,869
Jimmy Reece	18	Hoyt Machine No.5	Fred & Richard Sommer	Kurtis 500C	Offy	6	182	Broken throttle	2,749
Don Edmunds	19	McKay No.92	Roy McKay	Kurtis 500G	Offy	27	170	Spun out SE	3,169
Fred Agabashian	22	Bowes Seal Fast No.14	George Bignotti	Kurtis 500G	Offy	4	107	Oil leak	2,879
Mike Magill	24	Dayton Steel No.77	George Walther	Kurtis 500G	Offy	18	101	Wrecked MS	2,519
Eddie Johnson	25	Chapman No.43	H. A. Chapman	Kurtis 500G	Offy	20	93	Front wheel bearing	2,489
Bill Cheesbourg	26	Seal Line No.31	J. S. Donaldson	Kurtis 500G	Offy	23	81	Fuel leak	3,209
Al Keller	27	BardahlClancy No.16	Pat Clancy	Kurtis 500G	Offy	8	75	Wrecked SW	2,429
Jimmy Daywalt	28	Helse No.57	H. H. Johnson	Kurtis 500C	Offy	29	53	Wrecked NE	2,449
Ed Elisian	29	McNamara No.83	Kalamazoo Sports, Inc	Kurtis 500C	Offy	7	51	Broken timing gear	2,369
Eddie Russo	32	Sclavi & Amos No.55	Fred Sclavi	Kurtis 500C	Offy	26	0	Wrecked pacelap, BS	2,404
Elmer George	33	Travelon Trailer No.23	Ernest L. Ruiz	Kurtis 500B	Offy	9	0	Wrecked pacelap, BS	2,249
Dick Rathmann	99	Sumar No.18	Chapman S. Root	Kurtis 500G	Offy	17	0	Driven by Parsons	0
Billy Garrett	99	Federal Engineering No.33	Federal Automotive Assoc.	Kurtis 500E	Offy		0	DNS, too slow	0
Chuck Weyant	99	Jim Robbins No.35	Jim Robbins	Kurtis 500G	Offy		0	DNS, too slow	0
Ray Crawford	99	Meguiar Mirror Glaze No.49	Ray Crawford	Kurtis 500G	Offy		0	DNS, too slow	0
Bill Cheesbourg	99	Las Vegas Club No.45	Fred Gerhardt	Kurtis 500G	Offy		0	DNS, too slow	0
Jimmy Davies	99	Trio Brdeact wind allass No.32	Carl L. Anderson	Kurtis 500D	Offy		0	DNS, too slow	0
Don Edmunds	99	Braund Plywood No.67	Braund Plywood Co.	Kurtis 500C	Offy		0	DNS, too slow	0
Tony Bonadies	99	Ray Brady No.15	Ray T. Brady	Kurtis 500C	Offy		0	Wrecked practice, NW	0
Cliff Griffith	99	Seal Line No.31	J. S. Donaldson	Kurtis 500G	Offy		0	Did not qualify	0
Jim McWithey	99	Federal Engineering No.34	Federal Automotive Assoc.	Kurtis 500D	Offy		0	Wrecked qual, NE	0
Keith Andrews	99	Farina No.62	Giuseppe Farina	Kurtis 500G	Offy		0	Wrecked, fatal, NW	0
Johnny Fedricks	99	Gdula No.71	Edward Gdula	Kurtis 500C	Offy		0	Did not qualify	0
Langhorne, PA — 100 Mile — 2 Jun									
Art Bisch	3	Whittington No.37		Kurtis 4000	Offy	7	99	Flagged	2,040
Billy Garrett	8	Federal Engineering No.14	Federal Automotive Assoc.	Kurtis 4000	Offy	9	97	Flagged	688
Don Branson	9	Ray Brady No.15	Ray T. Brady	Kurtis 4000	Offy	5	96	Flagged	635
Van Johnson	10	Sam Traylor No.55	Samuel Traylor	Kurtis 4000	Offy	11	95	Flagged	583
Milwaukee, WI — 100 Mile — 9 Jun									
Johnny Boyd	3	Bowes Seal Fast No.6	George Bignotti	Kurtis 500G	Offy	4	100	Finished	2,149
Pat O'Connor	4	Sumar No.18	Chapman S. Root	Kurtis 500G	Offy	10	100	Finished	1,577
Jim McWithey	9	Federal Engineering No.33	Federal Automotive Assoc.	Kurtis 500E	Offy	11	99	Flagged	673
Bill Cheesbourg	11	Las Vegas Club No.45	Fred Gerhardt	Kurtis 500G	Offy	15	98	Flagged*	563
Jimmy Daywalt	13	Seal Line No.31	J. S. Donaldson	Kurtis 500G	Offy	13	96	Flagged	407
Fred Agabashian	17	Bowes Seal Fast No.14	George Bignotti	Kurtis 500G	Offy	12	84	Broken valve	181
Art Bisch	20	Whittington No.37		Kurtis 4000	Offy	20	59	Wrecked turn 3	136
Don Branson	21	Ray Brady No.15	Ray T. Brady	Kurtis 4000D	Offy	18	51	Stalled	90
Jack Turner	22	Bardahl No.19	Pat Clancy	Kurtis 500G	Offy	9	40	Lost brakes	90
Detroit, MI — 100 Mile — 23 Jun									
Billy Garrett	7	Federal Engineering No.33	Federal Automotive Assoc.	Kurtis 4000	Offy	9	100	Finished	460
Don Branson	10	Ray Brady No.15	Ray T. Brady	Kurtis 4000	Offy	8	98	Flagged	355
Chuck Weyant	11	Mel Wiggers No.35	Mel Wiggers	Kurtis 4000	Offy	16	97	Flagged	320
Bill Cheesbourg	12	Seal Line No.31	J. S. Donaldson	Kurtis 500G	Offy	15	95	Flagged	260
Ed Elisian	13	Lutes Truck Parts No.25	Francis Bardazon	Kurtis 4000	Offy	18	89	Spun out, brkn shock	225
Gene Hartley	15	Hart Fullerton No.22	Hart Fullerton	Kurtis 4000	Offy	13	56	Overheated	175
Van Johnson	99	Sam Traylor No.55	Samuel Traylor	Kurtis 4000	Offy		0	DNQ, too slow	0
Atlanta, GA — 100 Mile — 4 Jul									
Ed Elisian	5	Lutes Truck Parts No.25	Francis Bardazon	Kurtis 4000	Offy	12	100	Finished	650
Gene Hartley	7	Hart Fullerton No.22	Hart Fullerton	Kurtis 4000	Offy	11	100	Finished	360
Don Branson	8	Ray Brady No.15	Ray T. Brady	Kurtis 4000	Offy	6	100	Finished	331
Bill Cheesbourg	9	Seal Line No.31	J. S. Donaldson	Kurtis 500G	Offy	15	97	Flagged	303
Al Keller	12	Sam Traylor No.55	Samuel Traylor	Kurtis 4000	Offy	13	94	Spun out, turn 1	197

Driver	Fin Pos	Car Name	Entrant	Chassis	Engine	Start Pos	Laps	Time/Speed/Out	Purse
Springfield, IL — 100 Mile — 17 Aug									
Johnny Boyd	4	Bowes Seal Fast No.6	George Bignotti	Kurtis 500G	Offy	7	100	Finished	1,273
Ralph Liguori	10	Ray Brady No.15	Ray T. Brady	Kurtis 4000	Offy	17	98	Flagged	556
Mike Magill	13	Dayton Steel No.77	George Walther	Kurtis 4000	Offy	18	97	Flagged	364
Ed Elisian	15	Lutes Truck Parts No.17	Francis Bardazon	Kurtis 4000	Offy	15	79	Flagged	283
Van Johnson	16	Sam Traylor No.55	Samuel Traylor	Kurtis 4000	Offy	8	68	Poor handling	243
Jack Turner	99	Hart Fullerton No.22	Hart Fullerton	Kurtis 4000	Offy		0	DNS, too slow	0
Billy Garrett	99	Federal Engineerring No.33	Federal Automotive Assoc.	Kurtis 4000	Offy		0	DNS, too slow	0
Luther Baughman	99	Norris Ford No.85		Kurtis 500A	Pont		0	DNS, too slow	0
Milwaukee, WI — 200 Mile — 25 Aug									
Johnnie Parsons	7	Las Vegas Club No.45	Fred Gerhardt	Kurtis 500G	Offy	17	198	Flagged	1,133
Andy Linden	10	McNamara No.73	Kalamazoo Sports, Inc	Kurtis 500G	Offy	1	197	Flagged	983
			Lee Elkins						
Jack Turner	11	Hart Fullerton No.22	Hart Fullerton	Kurtis 4000D	Offy	23	196	Flagged	800
Don Freeland	15	Hoyt Machine No.5	Fred & Richard Sommer	Kurtis 500C	Offy	21	193	Flagged	380
Shorty Templeman	17	Pete Wales No.29	Peter Wales	Kurtis 4000D	Offy	11	151	Broken LF axle	303
Ed Elisian	20	Lutes Truck Parts No.17	Francis Bardazon	Kurtis 4000D	Offy	13	121	Engine failure	190
Johnny Boyd	24	Bowes Seal Fast No.6	George Bignotti	Kurtis 500G	Offy	4	38	Brkn universal joint	127
Bill Cheesbourg	99	Seal Line No.31	J. S. Donaldson	Kurtis 500G	Offy		0	DNS, too slow	0
Van Johnson	99	Sam Traylor No.55	Samuel Traylor	Kurtis 4000D	Offy		0	DNS, too slow	0
Mike Magill	99	Dayton Steel No.77	George Walther	Kurtis 500G	Offy		0	DNS, too slow	0
Ray Crawford	99	Crawford No.49	Ray Crawford	Kurtis 500G	Offy		0	DNS, too slow	0
Ralph Liguori	99	Ray Brady No.15	Ray T. Brady	Kurtis 4000D	Offy		0	DNS, too slow	0
Chuck Weyant	99	Mel Wiggers No.35	Mel Wiggers	Kurtis 4000D	Offy		0	DNS, too slow	0
Jim McWithey	99	Federal Engineering No.33	Federal Automotive Assoc.	Kurtis 500E	Offy		0	Wrecked practice	0
DuQuoin, IL — 100 Mile — 2 Sep									
Johnny Boyd	6	Bowes Seal Fast No.6	George Bignotti	Kurtis 500G	Offy	9	100	Finished	851
Ed Elisian	10	Lutes Truck Parts No.17	Francis Bardazon	Kurtis 4000	Offy	7	98	Flagged	531
Jack Turner	11	Hart Fullerton No.22	Hart Fullerton	Kurtis 4000	Offy	14	98	Flagged	483
Billy Garrett	12	Federal Engineering No.33	Federal Automotive Assoc.	Kurtis 4000	Offy	15	97	Flagged	396
Jimmy Reece	14	Sam Traylor No.55	Samuel Traylor	Kurtis 4000	Offy	13	95	Flagged	309
Ralph Liguori	17	Ray Brady No.15	Ray T. Brady	Kurtis 4000	Offy	18	32	Overheated	154
Syracuse, NY — 100 Mile — 7 Sep									
Johnny Boyd	9	Bowes Seal Fast No.6	George Bignotti	Kurtis 500G	Offy	15	98	Flagged	565
Jimmy Reece	10	Holynski No.96		Kurtis	Offy	16	98	Flagged	518
Ed Elisian	13	Lutes Truck Parts No.17	Francis Bardazon	Kurtis 4000	Offy	10	98	Flagged	337
Van Johnson	15	Sam Traylor No.55	Samuel Traylor	Kurtis 4000	Offy	12	98	Flagged	262
Mike Magill	16	Dayton Steel No.77	George Walther	Kurtis 4000	Offy	14	94	Flagged	225
Al Herman	99	Pete Wales No.29	Peter Wales	Kurtis 4000	Offy		0	DNS, too slow	0
Tony Bettenhausen	99	Federal Engineering No.33	Federal Automotive Assoc.	Kurtis 4000	Offy		0	DNS, too slow	0
Indianapolis, IN — 100 Mile — 14 Sep									
Ed Elisian	7	Lutes Truck Parts No.17	Francis Bardazon	Kurtis 4000	Offy	15	98	Flagged	1,148
Jack Turner	9	Hart Fullerton No.22	Hart Fullerton	Kurtis 4000	Offy	14	97	Flagged	992
Tony Bettenhausen	11	Federal Engineering No.33	Federal Automotive Assoc.	Kurtis 4000	Offy	17	96	Flagged	836
Johnny Boyd	14	Bowes Seal Fast No.6	George Bignotti	Kurtis 500G	Offy	6	77	Overheated	544
Mike Magill	99	Dayton Steel No.77	George Walther	Kurtis 4000	Offy		0	DNS, too slow	0
Van Johnson	99	Sam Traylor No.55	Samuel Traylor	Kurtis 4000	Offy		0	DNS, too slow	0
Shorty Templeman	99	Hardwood Door No.35	Ashley Wright	Kurtis 4000	Offy		0	DNS, too slow	0
Al Herman	99	Pete Wales No.29	Peter Wales	Kurtis 4000	Offy		0	DNS, too slow	0
Luther Baughman	99	Norris Ford No.85		Kurtis 500A	Pont		0	DNS, too slow	0
Bill Cheesbourg	99	Seal Line No.31	J. S. Donaldson	Kurtis 500G	Offy		0	Did not qualify	0
Trenton, NJ — 100 Mile — 29 Sep									
Johnny Boyd	9	Bowes Seal Fast No.6	George Bignotti	Kurtis 500G	Offy	19	98	Flagged	565
Ed Elisian	10	Lutes Truck Parts No.17	Francis Bardazon	Kurtis 4000D	Offy	14	98	Flagged	518
Andy Linden	13	McNamara No.73	Kalamazoo Sports, Inc	Kurtis 500G	Offy	20	97	Flagged	338
			Lee Elkins						
Jimmy Daywalt	14	Seal Line No.31	J. S. Donaldson	Kurtis 500G	Offy	16	97	Flagged	300
Shorty Templeman	17	Hardwood Door No.35	Ashley Wright	Kurtis 4000D	Offy	22	93	Flagged	150
Chuck Rodee	18	Pete Wales No.29	Peter Wales	Kurtis 4000D	Offy	21	87	Flagged	150
Mike Magill	99	Dayton Steel No.77	George Walther	Kurtis 4000D	Offy		0	DNS, too slow	0
Van Johnson	99	Sam Traylor No.55	Samuel Traylor	Kurtis 4000D	Offy		0	DNS, too slow	0
Sacramento, CA — 100 Mile — 20 Oct									
Johnny Boyd	6	Bowes Seal Fast No.6	George Bignotti	Kurtis 500G	Offy	4	98	Flagged	591
Jim Rathmann	12	Crawford No.49	Ray Crawford	Kurtis 4000	Offy	12	94	Flagged	266
Jimmy Reece	13	Dobry No.63	Leo Dobry	Kurtis 2000	Offy	15	94	Flagged	230
Ed Elisian	17	Hart Fullerton No.22	Hart Fullerton	Kurtis 4000	Offy	9	30	Magneto	102
Tony Bettenhausen	18	Las Vegas Club No.45	Fred Gerhardt	Kurtis 500G	Offy	8	0	Wrecked	102
R.D. Whittington	99	Whittington No.37		Kurtis 4000	Offy		0	DNS, too slow	0

Driver	Fin Pos	Car Name	Entrant	Chassis	Engine	Start Pos	Laps	Time/Speed/Out	Purse
Phoenix, AZ — 100 Mile — 11 Nov									
Johnny Boyd	3	Bowes Seal Fast No.6	George Bignotti	Kurtis 500G	Offy	2	100	Finished	2,173
Jim Rathmann	6	Crawford No.49	Ray Crawford	Kurtis 4000	Offy	10	99	Flagged	717
Earl Motter	9	Hart Fullerton No.22	Hart Fullerton	Kurtis 4000	Offy	11	97	Flagged	486
Del Fanning	99	Dobry No.63	Leo Dobry	Kurtis 2000	Offy		0	DNS, too slow	0
R.D. Whittington	99	Whittington No.37		Kurtis 4000	Offy		0	DNS, too slow	0
Monza, Italy — 166 Mile — 29 Jun									
Pat O'Connor	2	Sumar No.12	Chapman S. Root	Kurtis 500G	Offy	2	63	Finished	7,636
Andy Linden	3	McNamara Veedol No.73	Kalamazoo Sports, Inc Lee Elkins	Kurtis 500G	Offy	3	63	Finished	2,150
Eddie Sachs	4	Jim Robbins No.35	Jim Robbins	Kurtis 500G	Offy	5	62	Flagged	7,527
Ray Crawford	7	Meguiar Mirror Glaze No.49	Ray Crawford	Kurtis 500G	Offy	7	58	Flagged	2,227
Tony Bettenhausen	12	Novi Auto Air Condition No.27	Novi Racing Cars, Inc	Kurtis 500F	NoviSc	1	45	Broken sway bar	1,157
Paul Russo	99	Novi Air Conditioning No.54	Novi Racing Cars, Inc	Kurtis 500F	NoviSc	0		Brkn flywheel pract	0
Monza, Italy — 166 Mile — 29 Jun									
Ray Crawford	4	Meguiar Mirror Glaze No.49	Ray Crawford	Kurtis 500G	Offy	7	59	Flagged	0
Eddie Sachs	8	Jim Robbins No.35	Jim Robbins	Kurtis 500G	Offy	4	45	Brkn cam house bolts	0
Andy Linden	9	McNamara Veedol No.73	Kalamazoo Sports, Inc Lee Elkins	Kurtis 500G	Offy	3	27	Cracked frame	0
Pat O'Connor	10	Sumar No.12	Chapman S. Root	Kurtis 500G	Offy	2	16	Split fuel tank	0
Monza, Italy — 166 Mile — 29 Jun									
Pat O'Connor	7	Sumar No.12	Chapman S. Root	Kurtis 500G	Offy	8	9	Split fuel tank	0
Ray Crawford	8	Meguiar Mirror Glaze No.49	Ray Crawford	Kurtis 500G	Offy	4	1	Out	0
Williams Grove, PA — 25 Mile — 21 Jul									
Don Branson	5	Ray Brady No.15	Ray T. Brady	Kurtis 4000	Offy	3	50	Finished	525
Mike Magill	8	Dayton Steel No.77	George Walther	Kurtis 4000	Offy	10	50	Finished	487
Chuck Weyant	10	Mel Wiggers No.35	Mel Wiggers	Kurtis 4000	Offy	19	50	Finished	337
Van Johnson	13	Sam Traylor No.55	Samuel Traylor	Kurtis 4000	Offy	8	23	Out	112
Jack Turner	99	Hart Fullerton No.22	Hart Fullerton	Kurtis 4000	Offy	4	0	Did not start	75
Ed Elisian	99	Lutes Truck Parts No.17	Francis Bardazon	Kurtis 4000	Offy	9	0	Did not start	75
Charles Musselman	99	Peter Wales No.29	Ray T. Brady	Kurtis 4000	Offy	15	0	Did not start	131
Roy Graham	99	Holynski No.96		Kurtis	Offy	16	0	Did not start	112
1958									
Trenton, NJ — 100 Mile — 30 Mar									
Tony Bettenhausen	2	Hardwood Door No.35	Ashley Wright	Kurtis 4000D	Offy	5	100	Finished	2,660
Chuck Rodee	8	Holynski No.96		Kurtis D	Offy	10	72	Wrecked mainstretch	613
Red Amick	14	Dayton Steel No.77	George Walther, Jr	Kurtis 4000D	Offy	12	0	DNS, hit by towtruck	75
Jimmy Reece	99	Chiropractic No.16	Ray T. Brady	Kurtis 4000D	Offy		0	Clutch, practice	0
Indianapolis, IN — 500 Mile — 30 May									
Johnny Boyd	3	Bowes Seal Fast No.9	Robert M. Bowes II	Kurtis 500G	Offy	8	200	3:45:23.75 133.099	24,999
Eddie Johnson	9	Bryant Heating No.61	Duke Donaldson	Kurtis 500G	Offy	26	200	3:50:29.58 130.156	5,999
Bill Cheesbourg	10	Novi Auto Air Conditioner No.54	Novi Racing Corp., Inc	Kurtis 500F	NoviSc	33	200	3:52:17.35 129.149	6,399
Al Keller	11	Bardahl/Clancy No.52	Pat Clancy	Kurtis 500G	Offy	21	200	3:53:29.00 128.498	3,919
Johnnie Parsons	12	Gerhardt No.45	Fred Gerhardt	Kurtis 500G	Offy	6	200	3:53:54.60 128.254	3,599
Bob Christie	14	Federal Engineering No.65	Federal Automotive Associates	Kurtis 500E	Offy	17	189	Spun out, NE	4,209
Mike Magill	17	Dayton Steel No.77	George Walther, Jr	Kurtis 500G	Offy	31	136	Flagged	2,849
Paul Russo	18	Novi Auto Air Conditioner No.15	Novi Racing Corp., Inc	Kurtis 500F	NoviSc	14	122	Radiator leak	2,779
Shorty Templeman	19	McNamara No.83	Kalamazoo Sports, Inc	Kurtis 500C	Offy	23	116	Lost brakes	2,699
Billy Garrett	21	Chapman No.43	H.A. Chapman	Kurtis 500G	Offy	15	80	Broken cam gear	2,639
Johnny Thomson	23	D-A Lubricants No.7	Racing Associates	Kurtis	Offy	22	52	Steering damaged	2,754
Bob Veith	26	Bowes Seal Fast No.14	Robert M. Bowes II	Kurtis 500G	Offy	4	1	Wrecked NE	2,789
Pat O'Connor	29	Sumar No.4	Chapman S. Root	Kurtis 500G	Offy	5	0	Wrecked NE, fatal	2,574
Paul Goldsmith	30	City of Daytona Beach No.31	Smokey Yunick	Kurtis 500G	Offy	16	0	Wrecked NE	2,369
Jerry Unser	31	McKay Bulldog No.92	Roy McKay	Kurtis 500G	Offy	24	0	Wrecked NE	2,339
Len Sutton	32	Jim Robbins No.68	Jim Robbins	Kurtis 500G	Offy	27	0	Wrecked NE	2,309
Gene Hartley	99	Hoyt Machine No.24	F. W. & R. F. Sommer	Kurtis 500C	Offy		0	DNS, too slow	0
Fred Agabashian	99	City of Memphis No.56	Pat Clancy	Kurtis 500G	Offy		0	DNS, too slow	0
Ray Crawford	99	Meguiar Mirror Glaze No.49	Ray Crawford	Kurtis 500G	Offy		0	DNS, too slow	0
Art Bisch	99	Ansted Rotary No.17	Jack R. Ensley	Kurtis 500D	Offy		0	DNS, too slow	0
Rex Easton	99	Hoover Motor Express No.10	John R. Wills	Kurtis 500G	Offy		0	DNS, too slow	0
Jack Ensley	99	Ansted Rotary No.17	Jack R. Ensley	Kurtis 500D	Offy		0	Did not qualify	0
Marshall Teague	99	Sumar No.18	Chapman S. Root	Kurtis 500G	Offy		0	Did not qualify	0
Jerry Unser	99	Sumar No.48	Chapman S. Root	Kurtis 500C	Offy		0	Did not qualify	0
Jerry Unser	99	Duncan No.51	H. Duncan	Kurtis 500A	DeStV8	0		Did not qualify	0
Juan Fangio	99	Novi Auto Air Conditioner No.54	Novi Racing Corp., Inc	Kurtis 500F	NoviSc	0		Did not qualify	0
Van Johnson	99	Ray Brady Chiropractic No.58	Ray T. Brady	Kurtis 500C	Offy		0	Wrecked practice, NE	0
Joe Giba	99	Carl Anderson No.62	Carl L. Anderson	Kurtis 500D	Offy		0	Did not qualify	0
Jimmy Daywalt	99	Federal Engineering No.69	Federal Automotive	Kurtis 500D	Offy		0	Did not qualify	0

Driver	Fin Pos	Car Name	Entrant	Chassis	Engine	Start Pos	Laps	Time/Speed/Out	Purse
			Associates						
Rex Easton	99	Wyandotte Tool No.72	John Marco Pusilo	Kurtis 500A	Offy		0	Did not qualify	0
Bill Homeier	99	Kurtis-Kraft/Smith No.74		Kurtis 500C	Offy		0	Did not qualify	0
Juan Fangio	99	Dayton Steel No.77	George Walther, Jr	Kurtis 500G	Offy		0	Did not qualify	0
Bob Cortner	99	McKay Bulldog No.93	Roy McKay	Kurtis 3000	Offy		0	Did not qualify	0
LeRoy Warriner	99	Safety Auto Glass No.95	James Edward Shreve	Kurtis 500C	Offy		0	Did not qualify	0
Bill Homeier	99	Safety Auto Glass No.95	James Edward Shreve	Kurtis 500C	Offy		0	Did not qualify	0
Milwaukee, WI — 100 Mile — 8 Jun									
Tony Bettenhausen	2	Hardwood Door No.35	Ashley Wright	Kurtis 4000D	Offy	3	100	Finished	2,780
Johnny Boyd	5	Bowes Seal Fast No.14	Robert M. Bowes II	Kurtis 500G	Offy	7	100	Finished	1,148
Bill Cheesbourg	12	Chiropractic No.16	Ray T. Brady	Kurtis 4000D	Offy	22	98	Flagged	605
Rex Easton	15	Wyandotte Tool No.72	John Marco Pusilo	Kurtis 500A	Offy	14	98	Spun Out, turn 1	237
Earl Motter	16	City Of Memphis No.56	Pat Clancy	Kurtis 500G	Offy	19	96	Broken steering	197
Billy Garrett	18	Chapman No.43	H.A. Chapman	Kurtis 500G	Offy	2	53	Wrecked, turn 3	158
Eddie Johnson	22	Bryant Heating No.61	Duke Donaldson	Kurtis 500G	Offy	15	2	Spun out, turn 4	79
Bob Veith	99	Bowes Seal Fast No.14	Robert M. Bowes II	Kurtis 500G	Offy		0	DNS, too slow	0
Jack Turner	99	Hart Fullerton No.25	Hart Fullerton	Kurtis 4000D	Offy		0	DNS, too slow	20
Don Branson	99	Hoover Motor Express No.10	John R. Wills	Kurtis 500G	Offy		0	Did not qualify	20
Ed Elisian	99	Bryant Heating No.61	Duke Donaldson	Kurtis 500G	Offy		0	Did not qualify	0
Langhorne, PA — 100 Mile — 15 Jun									
Tony Bettenhausen	5	Hardwood Door No.35	Ashley Wright	Kurtis 4000	Offy	7	100	Finished	1,100
Mike Magill	7	Vargo No.53	Jake Vargo	Kurtis 4000	Offy	13	100	Finished	660
Bill Cheesbourg	10	Chiropractic No.16	Ray T. Brady	Kurtis 4000	Offy	17	98	Flagged	518
Jack Turner	12	Hart Fullerton No.25	Hart Fullerton	Kurtis 4000	Offy	9	90	Wrecked, backstretch	385
Rex Easton	17	Wyandotte Tool No.72	John Marco Pusilo	Kurtis 4000	Offy	12	9	Driver tired	150
Earl Motter	99	Peter Wales Trucking No.15	Peter Wales	Kurtis 4000	Offy		0	DNS, too slow	0
Atlanta, GA — 100 Mile — 4 Jul									
Bill Cheesbourg	10	Chiropractic No.16	Ray T. Brady	Kurtis 4000	Offy	13	98	Flagged	298
Roy Graham	12	Westwood Gage No.15	Peter Wales	Kurtis 4000	Offy	18	97	Flagged	216
Rex Easton	13	Wyandotte Tool No.72	John Marco Pusilo	Kurtis 4000	Offy	15	96	Flagged	186
Jack Turner	17	Hart Fullerton No.25	Hart Fullerton	Kurtis 4000	Offy	11	37	Stopped	82
Tony Bettenhausen	18	Hardwood Door No.35	Ashley Wright	Kurtis 4000	Offy	7	14	Magneto	82
Springfield, IL — 100 Mile — 16 Aug									
Rex Easton	5	Wyandotte Tool No.72	John Marco Pusilo	Kurtis 4000	Offy	6	99	Flagged	1,163
Earl Motter	10	Hart Fullerton No.25	Hart Fullerton	Kurtis 4000	Offy	7	87	Out of fuel	552
Tony Bettenhausen	12	Hardwood Door No.35	Ashley Wright	Kurtis 4000	Offy	10	82	Mechanical	411
Bill Cheesbourg	99	Chiropractic No.16	Ray T. Brady	Kurtis 4000	Offy		0	DNS, too slow	0
Van Johnson	99	Vargo No.53	Jake Vargo	Kurtis 4000	Offy		0	DNS, too slow	0
Roy Graham	99	Westwood Gage No.15	Peter Wales	Kurtis 4000	Offy		0	DNS, too slow	0
Milwaukee, WI — 200 Mile — 24 Aug									
Bob Veith	2	Bowes Seal Fast No.14	Robert M. Bowes II	Kurtis 500G	Offy	2	200	Finished	4,380
Jack Turner	5	Bowes Seal Fast No.9	Robert M. Bowes II	Kurtis 500G	Offy	7	199	Flagged	1,761
Eddie Johnson	10	Bryant Heating No.61	Duke Donaldson	Kurtis 500G	Offy	12	198	Flagged	885
Shorty Templeman	13	Westwood Gage No.15	Peter Wales	Kurtis 4000D	Offy	15	189	Flagged	572
Earl Motter	14	Hart Fullerton No.25	Hart Fullerton	Kurtis 4000D	Offy	26	188	Flagged	508
Rex Easton	17	Wyandotte Tool No.72	John Marco Pusilo	Kurtis 500A	Offy	25	178	Flagged	254
Jimmy Davies	20	Chiropractic No.16	Ray T. Brady	Kurtis 4000D	Offy	17	141	Out of fuel	190
Tony Bettenhausen	25	Hardwood Door No.35	Ashley Wright	Kurtis 4000D	Offy	14	36	Ignition trouble	127
Bill Cheesbourg	99	Federal Engineering No.65	Federal Automotive Assoc..	Kurtis 500E	Offy		0	DNS, too slow	0
Tony Bonadies	99	Tassi Vatis No.49	Tassi Vatis	Kurtis 4000D	Offy		0	DNS, too slow	0
DuQuoin, IL — 100 Mile — 1 Sep									
Rex Easton	10	Wyandotte Tool No.72	John Marco Pusilo	Kurtis 4000	Offy	12	99	Flagged	622
Shorty Templeman	11	Westwood Gage No.15	Peter Wales	Kurtis 4000	Offy	15	99	Flagged	566
Dick Linder	14	Vargo No.53	Jake Vargo	Kurtis 4000	Offy	16	90	Flagged	364
Jimmy Davies	15	Chiropractic No.16	Ray T. Brady	Kurtis 4000	Offy	18	62	Broken rear end	318
Earl Motter	16	Hart Fullerton No.25	Hart Fullerton	Kurtis 4000	Offy	7	60	Spun ou, turn 4	273
Bill Chennault	99	Hardwood Door No.35	Ashley Wright	Kurtis 4000	Offy		0	DNS, too slow	0
Syracuse, NY — 100 Mile — 6 Sep									
Rex Easton	7	Wyandotte Tool No.72	John Marco Pusilo	Kurtis 4000	Offy	11	99	Flagged	660
Jack Turner	11	Hardwood Door No.35	Ashley Wright	Kurtis 4000	Offy	17	98	Flagged	470
Shorty Templeman	12	Westwood Gage No.15	Peter Wales	Kurtis 4000	Offy	18	98	Flagged	385
Dick Linder	17	Vargo No.53	Jake Vargo	Kurtis 4000	Offy	12	50	Engine	150
Earl Motter	99	Holynski No.96		Kurtis	Offy		0	DNS, too slow	0
Indianapolis, IN — 100 Mile — 13 Sep									
Earl Motter	6	Hart Fullerton No.25	Hart Fullerton	Kurtis 4000	Offy	8	99	Flagged	1,490
Jimmy Reece	7	Bowes Seal Fast No.14	Robert M. Bowes II	Kurtis 500G	Offy	7	99	Flagged	1,188
Rex Easton	99	Wyandotte Tool No.72	John Marco Pusilo	Kurtis 4000	Offy		0	DNS, too slow	0

Driver	Fin Pos	Car Name	Entrant	Chassis	Engine	Start Pos	Laps	Time/Speed/Out	Purse
Jimmy Davies	99	Chiropractic No.16	Ray T. Brady	Kurtis 4000	Offy		0	DNS, too slow	0
Shorty Templeman	99	Westwood Gage No.15	Peter Wales	Kurtis 4000	Offy		0	DNS, too slow	0
Eddie Johnson	99	Bryant Heating No.61	Duke Donaldson	Kurtis 500G	Offy		0	DNS, too slow	0
Jack Turner	99	Hardwood Door No.35	Ashley Wright	Kurtis 4000	Offy		0	DNS, too slow	0
Jimmy Daywalt	99	Dayton Steel No.77	George Walther, Jr	Kurtis 4000	Offy		0	DNS, too slow	0
Dick Linder	99	Vargo No.53	Jake Vargo	Kurtis 4000	Offy		0	DNS, too slow	0
Trenton, NJ — 100 Mile — 28 Sep									
Eddie Johnson	5	Bryant Heating No.61	Duke Donaldson	Kurtis 500G	Offy	5	99	Flagged	1,100
Jimmy Reece	7	Bowes Seal Fast No.14	Robert M. Bowes II	Kurtis 500G	Offy	1	99	Wrecked, T1 - fatal	660
Rex Easton	13	Wyandotte Tool No.72	John Marco Pusilo	Kurtis 4000D	Offy	12	96	Flagged	338
Bob Veith	16	Chiropractic No.16	Ray T. Brady	Kurtis 4000D	Offy	21	91	Flagged	187
Dick Linder	17	Vargo No.53	Jake Vargo	Kurtis 4000D	Offy	19	88	Flagged	150
Shorty Templeman	20	Westwood Gage No.15	Peter Wales	Kurtis 4000D	Offy	17	60	Mechanical trouble	112
Tony Bonadies	21	Tassi Vatis No.49	Tassi Vatis	Kurtis 4000D	Offy	22	14	Mechanical trouble	75
Jim McWithey	99	Federal Engineering No.65	Federal Automotive Assoc.	Kurtis 500E	Offy		0	DNS, too slow	0
Jack Turner	99	Hardwood Door No.35	Ashley Wright	Kurtis 4000D	Offy		0	DNS, too slow	0
Mike Magill	99	Dayton Steel No.77	George Walther, Jr	Kurtis 4000D	Offy		0	Did not qualify	0
Sacramento, CA — 100 Mile — 26 Oct									
Bob Veith	6	Chiropractic No.16	Ray T. Brady	Kurtis 4000	Offy	11	100	Finished	783
Wayne Weiler	9	Hart Fullerton No.25	Hart Fullerton	Kurtis 4000	Offy	13	96	Flagged	532
Don Branson	12	Hardwood Door No.35	Ashley Wright	Kurtis 4000	Offy	5	96	Flagged	362
Rex Easton	13	Wyandotte Tool No.72	John Marco Pusilo	Kurtis 4000	Offy	16	96	Flagged	316
Bill Hyde	99	Scovell No.65	Robert Scovell	Kurtis 2000	Offy		0	DNS, too slow	0
Ralph Liguori	99	Tassi Vatis No.49	Tassi Vatis	Kurtis 4000	Offy		0	DNS, too slow	0
Phoenix, AZ — 100 Mile — 11 Nov									
Rex Easton	6	Wyandotte Tool No.72	John Marco Pusilo	Kurtis 4000	Offy	1	98	Flagged	685
Lloyd Ruby	13	Hardwood Door No.35	Ashley Wright	Kurtis 4000	Offy	16	93	Flagged	272
Bill Homeier	16	Tassi Vatis No.49	Tassi Vatis	Kurtis 4000	Offy	8	46	Out	181
Wayne Weiler	17	Hart Fullerton No.25	Hart Fullerton	Kurtis 4000	Offy	11	34	Broken RR axle	121
Bob Veith	18	Chiropractic No.16	Ray T. Brady	Kurtis 4000	Offy	18	24	Poor handling	121
Bill Hyde	99	Scovell No.65	Robert Scovell	Kurtis 2000	Offy		0	DNS, too slow	0
Bill Cheesbourg	99	Peter Wales Trucking No.15	Peter Wales	Kurtis 4000	Offy		0	DNS, too slow	0
R.D. Whittington	99	Whittington No.37		Kurtis 4000	Offy		0	DNS, too slow	0
Monza, Italy — 166 Mile — 29 Jun									
Bob Veith	3	Bowes Seal Fast No.9	Robert M. Bowes II	Kurtis 500G	Offy	2	62	Flagged	4,120
Jimmy Reece	8	Hoyt Machine No.24	F. W. & R. F. Sommer	Kurtis 500C	Offy	13	59	Flagged	3,041
Ray Crawford	10	Meguiar's Mirror Glaze No.49	Ray Crawford	Kurtis 500G	Offy	12	58	Flagged	3,677
Eddie Sachs	15	Jim Robbins No.35	Jim Robbins	Kurtis 500G	Offy	4	20	Brkn connecting rod	1,761
Monza, Italy — 166 Mile — 29 Jun									
Bob Veith	2	Bowes Seal Fast No.9	Robert M. Bowes II	Kurtis 500G	Offy	3	63	Finished	0
Jimmy Reece	7	Hoyt Machine No.24	F. W. & R. F. Sommer	Kurtis 500C	Offy	8	60	Flagged	0
Ray Crawford	8	Meguiar's Mirror Glaze No.49	Ray Crawford	Kurtis 500G	Offy	10	60	Flagged	0
Monza, Italy - 166 Mile - 29 Jun									
Ray Crawford	4	Meguiar's Mirror Glaze No.49	Ray Crawford	Kurtis 500G	Offy	8	60	Flagged	0
Jimmy Reece	5	Hoyt Machine No.24	F. W. & R. F. Sommer	Kurtis 500C	Offy	7	59	Flagged	0
Bob Veith	10	Bowes Seal Fast No.9	Robert M. Bowes II	Kurtis 500G	Offy	2	28	Lost wheel	0
Williams Grove, PA — 25 Mile — 20 Jul									
Rex Easton	9	Wyandotte Tool No.72	John Marco Pusilo	Kurtis 4000	Offy	10	50	Finished	263
Van Johnson	10	Vargo No.53	Jake Vargo	Kurtis 4000	Offy	15	50	Finished	338
Roy Graham	11	Peter Wales Trucking No.15	Peter Wales	Kurtis 4000	Offy	17	50	Finished	338
Earl Motter	12	Hart Fullerton No.25	Hart Fullerton	Kurtis 4000	Offy	7	33	Spun out, turn 4	300
Bill Cheesbourg	99	Ray Brady No.16	Ray T. Brady	Kurtis 4000	Offy	16	0	Did not start	113
1959									
Daytona, FL — 100 Mile — 4 Apr									
Bob Christie	3	Federal Engineering No.65	Federal Automotive Assoc.	Kurtis 500D	Offy	6	39	Flagged	1,980
Dick Rathmann	5	Sumar No.41	Chapman S. Root	Kurtis 500G	Offy	1	39	Flagged	1,563
Bill Cheesbourg	12	Wright-Knight No.25	Ashley Wright	Kurtis 4000D	Offy	12	36	Spun to miss Amick	385
Bill Randall	14	Safety Auto Glass No.95	Shreve, Perkins, Greismer	Kurtis 500C	Offy	19	34	Flagged	300
Dempsey Wilson	17	Hoyt Machine No.24	Fred & Richard Sommer	Kurtis 500C	Offy	4	28	Spun out, turn 2	150
Trenton, NJ — 100 Mile — 19 Apr									
Van Johnson	10	Ray Brady No.59	Ray T. Brady	Kurtis 4000D	Offy	21	79	Flagged	518
Dick Linder	14	Vargo No.5	Jake Vargo	Kurtis 4000D	Offy	12	56	Wrecked, fatal, T3	300
Bob Christie	15	Federal Engineering No.65	Federal Automotive Assoc.	Kurtis 500D	Offy	11	56	Spun to miss Branson	225
Tony Bonadies	21	Tassi Vatis No.94	Tassi Vatis	Kurtis 4000D	Offy	16	15	Out	75
Bill Cheesbourg	99	Hardwood Door No.25	Ashley Wright	Kurtis 500G	Offy		0	DNQ, seized engine	38

Driver	Fin Pos	Car Name	Entrant	Chassis	Engine	Start Pos	Laps	Time/Speed/Out	Purse
Indianapolis, IN — 500 Mile — 30 May									
Duane Carter	7	Smokey's Reverse Torque No.37	Yunick, Glover & Lathrop	Kurtis	Offy	12	200	3:44:59.15 133.342	7,275
Eddie Johnson	8	Bryant Heating & Air Cond No.19		Kurtis 500G	Offy	8	200	3:44:59.69 133.336	6,625
Paul Russo	9	Bardahl\Gerhardt No.45		Kurtis 500G	Offy	27	200	3:45:00.24 133.331	6,325
Jimmy Daywalt	14	Federal Engineering No.66	Federal Automotive Assoc.	Kurtis 500E	Offy	13	200	3:47:30.72 131.861	5,335
Jim McWithey	16	Ray Brady No.58	Ray T. Brady	Kurtis 500C	Offy	33	200	3:52:30.93 129.204	4,045
Don Freeland	22	Jim Robbins No.15	Jim Robbins	Kurtis 500G	Offy	25	136	Broken valve spring	4,485
Bob Christie	25	Federal Engineering No.65	Federal Automotive Assoc.	Kurtis 500D	Offy	24	109	Broken rod bolt	4,595
Bobby Grim	26	Sumar No.48	Chapman S. Root	Kurtis 500G	Offy	5	85	Burned piston	4,190
Chuck Weyant	28	McKay No.47	Roy McKay	Kurtis 500J	Offy	29	45	Wrecked NE	3,505
Jud Larson	29	Bowes Seal Fast No.7	Bignotti-Bowes Racing	Kurtis 500J	Offy	19	45	Wrecked NE	3,650
Red Amick	31	Wheeler-Foutch No.87	LeRoy E. Foutch, Jr	Kurtis 500C	Offy	26	45	Wrecked NE	3,915
Bill Homeier	99	Rusco Battery Cable No.62		Kurtis 500G	Offy		0	DNS, too slow	0
Johnny Kay	99	Safety Auto Glass No.17	Shreve, Perkins, Greismer	Kurtis 500C	Offy		0	Did not qualify	0
Bill Cheesbourg	99	Hardwood Door No.25	Ashley Wright	Kurtis 500G	Offy		0	Did not qualify	0
Dempsey Wilson	99	Novi Diesel No.34	Novi Racing Corp., Inc	Kurtis 500F	NoviSc	0		Did not qualify	0
Paul Russo	99	Novi Diesel No.38	Novi Racing Corp., Inc	Kurtis 500F	NoviSc	0		Did not qualify	0
Ralph Liguori	99	Sumar No.41	Chapman S. Root	Kurtis 500G	Offy		0	Did not qualify	0
Jimmy Davies	99	Sumar No.41	Chapman S. Root	Kurtis 500G	Offy		0	Did not qualify	0
Bill Homeier	99	Go Kart No.42		Kurtis 500G	Offy		0	Did not qualify	0
Shorty Templeman	99	Frank Arciero No.69	Frank Arciero	Kurtis 500C	Maser	0		Did not qualify	0
Chuck Arnold	99	Speed Enterprise No.72	Leonard G. Smith	Kurtis 500C	Offy		0	Did not qualify	0
Bob Schroeder	99	Midwest Manufacturing No.78	Carl Lee Gehlhausen	Kurtis 500D	Offy		0	Did not qualify	0
Gene Force	99	Midwest Manufacturing No.78	Carl Lee Gehlhausen	Kurtis 500D	Offy		0	Did not qualify	0
Eddie Russo	99	Midwest Manufacturing No.78	Carl Lee Gehlhausen	Kurtis 500D	Offy		0	Did not qualify	0
Johnny Moorehouse	99	Binter No.91		Kurtis 500G	Offy		0	Wrecked practice	0
Jack Ensley	99	McKay No.92	Roy McKay	Kurtis 500G	Offy		0	Did not qualify	0
Eddie Russo	99	McKay No.93	Roy McKay	Kurtis 500G	Offy		0	Did not qualify	0
Milwaukee, WI — 100 Mile — 7 Jun									
Jud Larson	4	Bowes Seal Fast No.7	Bignotti-Bowes Racing	Kurtis 500G	Offy	5	100	Finished	1,859
Don Freeland	8	Jim Robbins No.15	Jim Robbins	Kurtis 500G	Offy	15	99	Flagged	869
Jimmy Daywalt	99	Bryant Heating & Air Cond No.19		Kurtis 500G	Offy		0	DNS, too slow	0
Jim McWithey	99	Ray Brady No.58	Ray T. Brady	Kurtis 500C	Offy		0	DNS, too slow	0
Bob Christie	99	Federal Engineering No.65	Federal Automotive Assoc.	Kurtis 500D	Offy		0	DNS, too slow	0
Dick Rathmann	99	Hardwood Door No.25	Ashley Wright	Kurtis 500G	Offy		0	DNS, too slow	0
Wayne Weiler	99	Fike Plumbing No.39	Harlan Fike	Kurtis 4000D	Offy		0	DNS, too slow	0
Paul Russo	99	Bardahl\Gerhardt No.45		Kurtis 500G	Offy		0	DNS, too slow	0
Langhorne, PA — 100 Mile — 14 Jun									
Van Johnson	1	Vargo No.56	Jake Vargo	Kurtis 4000	Offy	11	100	1:00:16.176 99.553	5,525
Ed Elisian	3	Hart Fullerton No.39	Hart Fullerton	Kurtis 4000	Offy	13	100	Finished	2,330
Dick Rathmann	7	Hardwood Door No.25	Ashley Wright	Kurtis 4000	Offy	12	100	Finished	800
Bill Homeier	99	Tassi Vatis No.94	Tassi Vatis	Kurtis 4000	Offy		0	DNS, too slow	0
Jimmy Davies	99	Peter Wales No.15	Peter Wales	Kurtis 4000	Offy		0	Did not qualify	0
Jim McWithey	99	Ray Brady No.59	Ray T. Brady	Kurtis 4000	Offy		0	Did not qualify	0
Springfield, IL — 100 Mile — 22 Aug									
Dick Rathmann	6	Hardwood Door No.25	Ashley Wright	Kurtis 4000	Offy	16	100	Finished	916
Wayne Weiler	10	Fike Plumbing No.39	Harlan Fike	Kurtis 4000	Offy	14	98	Flagged	573
Chuck Weyant	11	Wyandotte Tool No.72	John Pusilo	Kurtis 4000	Offy	17	97	Flagged	522
Dempsey Wilson	99	Federal Engineering No.34	Federal Automotive Assoc.	Kurtis 4000	Offy		0	DNS, too slow	0
Bill Homeier	99	Tassi Vatis No.94	Tassi Vatis	Kurtis 4000	Offy		0	DNS, too slow	0
Chuck Rodee	99	Ray Brady No.59	Ray T. Brady	Kurtis 4000	Offy		0	DNS, too slow	0
Herb Hill	99	Hill Carpets No.89	Herb Hill	Kurtis 500A	Stock	0		Did not qualify	0
Milwaukee, WI — 200 Mile — 30 Aug									
Don Freeland	3	Bowes Seal Fast No.27	Bignotti-Bowes Racing	Kurtis 500J	Offy	14	200	Finished	2,948
Dick Rathmann	7	Hardwood Door No.25	Ashley Wright	Kurtis 500G	Offy	3	200	Finished	1,155
Wayne Weiler	12	Fike Plumbing No.39	Harlan Fike	Kurtis 4000D	Offy	12	197	Flagged	667
Dempsey Wilson	13	Federal Engineering No.34	Federal Automotive Assoc.	Kurtis 4000D	Offy	23	195	Flagged	582
Chuck Arnold	18	Bryant Heating & Air Cond No.19		Kurtis 500G	Offy	25	129	Spun out	259
Eddie Johnson	23	Hardwood Door No.35	Ashley Wright	Kurtis 4000D	Offy	9	27	Wrecked, turn 2	129
Bobby Grim	26	Sumar No.41	Chapman S. Root	Kurtis 500G	Offy	24	13	Spun out	129
Paul Russo	99	Bardahl\Gerhardt No.45		Kurtis 500G	Offy		0	DNS, too slow	0
Cotton Farmer	99	McKay No.92	Roy McKay	Kurtis 500G	Offy		0	DNS, too slow	0
Herb Hill	99	Hill Carpets No.99	Herb Hill	Kurtis 500A	Offy		0	DNS, too slow	0
Red Amick	99	Wheeler-Foutch No.84		Kurtis 500C	Offy		0	Did not qualify	0
Bob Wente	99	Safety Auto Glass No.95	Shreve, Perkins, Greismer	Kurtis 500C	Offy		0	Could not start	0
DuQuoin, IL — 100 Mile — 7 Sep									
Chuck Weyant	11	Wyandotte Tool No.72	John Pusilo	Kurtis 4000	Offy	16	98	Flagged	566
Wayne Weiler	14	Fike Plumbing No.39	Harlan Fike	Kurtis 4000	Offy	14	96	Flagged	364
Shorty Templeman	17	Federal Engineering No.34	Federal Automotive Assoc.	Kurtis 4000	Offy	15	53	Wrecked	182

Driver	Fin Pos	Car Name	Entrant	Chassis	Engine	Start Pos	Laps	Time/Speed/Out	Purse
Chuck Weyant	99	Wyandotte Tool No.72	John Pusilo	Kurtis 4000	Offy		0	DNS, too slow	0
Bill Homeier	99	Tassi Vatis No.94	Tassi Vatis	Kurtis 4000	Offy		0	DNS, too slow	0
Syracuse, NY — 100 Mile — 12 Sep									
Bill Homeier	10	Tassi Vatis No.94	Tassi Vatis	Kurtis 4000	Offy	15	99	Flagged	518
Chuck Weyant	11	Wyandotte Tool No.72	John Pusilo	Kurtis 4000	Offy	14	99	Flagged	470
Shorty Templeman	13	Federal Engineering No.34	Federal Automotive Associates	Kurtis 4000	Offy	9	98	Flagged	337
Ralph Liguori	99	Ray Brady No.59	Ray T. Brady	Kurtis 4000	Offy		0	DNS, too slow	0
Indianapolis, IN — 100 Mile — 19 Sep									
Bill Homeier	9	Tassi Vatis No.94	Tassi Vatis	Kurtis 4000	Offy	18	98	Flagged	1,118
Shorty Templeman	12	Federal Engineering No.34	Federal Automotive Associates	Kurtis 4000	Offy	7	97	Flagged	780
Dempsey Wilson	99	Fike Plumbing No.39	Harlan Fike	Kurtis 4000	Offy		0	DNS, too slow	0
Dick Rathmann	99	Hardwood Door No.25	Ashley Wright	Kurtis 4000	Offy		0	DNS, too slow	0
Jimmy Davies	99	Lutes Truck Parts No.97		Kurtis 4000	Offy		0	DNS, too slow	0
Chuck Weyant	99	Wyandotte Tool No.72	John Pusilo	Kurtis 4000	Offy		0	DNS, too slow	0
Chuck Arnold	99	Ray Brady No.59	Ray T. Brady	Kurtis 4000	Offy		0	DNS, too slow	0
Johnny Kay	99	Safety Auto Glass No.95	Shreve, Perkins, Greismer	Kurtis 500C	Offy		0	DNS, too slow	0
Herb Hill	99	Hill Carpets No.63	Herb Hill	Kurtis 500A	Pont		0	Did not qualify	0
Trenton, NJ — 100 Mile — 27 Sep									
Eddie Johnson	3	Hardwood Door No.25	Ashley Wright	Kurtis 500G	Offy	9	100	Finished	1,980
Shorty Templeman	6	Federal Engineering No.34	Federal Automotive Associates	Kurtis 4000D	Offy	6	100	Finished	830
Chuck Arnold	13	Bryant Heating & Air Cond No.19		Kurtis 500G	Offy	16	95	Flagged	338
Tony Bonadies	16	Tassi Vatis No.94	Tassi Vatis	Kurtis 4000D	Offy	19	91	Flagged	188
Cotton Farmer	20	Wyandotte Tool No.72	John Pusilo	Kurtis 4000D	Offy	22	52	Black flag, too slow	112
George Morris	99	McConnell No.29		Kurtis 3000D	Offy		0	DNS, too slow	0
Phoenix, AZ — 100 Mile — 18 Oct									
Shorty Templeman	9	Federal Engineering No.34	Federal Automotive Assoc.	Kurtis 4000	Offy	12	98	Flagged	399
Jack Rounds	10	Fike Plumbing No..39	Harlan Fike	Kurtis 4000	Offy	5	97	Flagged	364
Bill Homeier	11	Tassi Vatis No.94	Tassi Vatis	Kurtis 4000	Offy	10	97	Flagged	328
Dempsey Wilson	12	Bardahl\Gerhardt No.45		Kurtis 500G	Offy	13	96	Flagged	267
Bill Hyde	99	Scovell No.65	Robert Scovell	Kurtis 2000	Offy		0	DNS, too slow	0
John Wood	99	Binter No.91		Kurtis 500G	Offy		0	DNS, too slow	0
Sacramento, CA — 100 Mile — 25 Oct									
Bill Homeier	5	Tassi Vatis No.94	Tassi Vatis	Kurtis 4000	Offy	7	75	Flagged	1,052
Shorty Templeman	6	Federal Engineering No.34	Federal Automotive Assoc.	Kurtis 4000	Offy	17	98	Flagged	790
Bill Hyde	9	Scovell No.65	Robert Scovell	Kurtis 2000	Offy	13	69	Out	537
Dempsey Wilson	16	Bardahl\Gerhardt No.45		Kurtis 500G	Offy	18	12	Out	213
A.J. Foyt	98	Tassi Vatis No.94	Tassi Vatis	Kurtis 4000	Offy		23	Rel Homeier 76-98	0
Jack Rounds	99	Fike Plumbing No.39	Harlan Fike	Kurtis 4000	Offy		0	DNS, mechanical	0
Dempsey Wilson	99	Bardahl\Gerhardt No.45		Kurtis 500G	Offy		0	DNS, too slow	0
Williams Grove, PA — 25 Mile — 19 Jul									
Chuck Arnold	8	Ray Brady No.59	Ray T. Brady	Kurtis 4000	Offy	12	13	Flagged - rain	337
Bill Homeier	9	Tassi Vatis No.94	Tassi Vatis	Kurtis 4000	Offy	9	13	Flagged - rain	337
Beau Clerke	11	McConnell No.29		Kurtis 3000	Offy	16	13	Flagged - rain	337
Van Johnson	12	Vargo No.56	Jake Vargo	Kurtis 4000	Offy	15	2	Wrecked,turn 3,fatal	300
Joe Barzda	13	Peter Wales No.15	Peter Wales	Kurtis 4000	Offy	13	2	Wrecked,turn 3	225
Ed Elisian	99	Fike Plumbing No.39	Harlan Fike	Kurtis 4000	Offy	6	0	Did not start	112
Daytona, FL — 50 Mile — 4 Apr									
Dick Rathmann	2	Sumar No.41	Chapman S. Root	Kurtis 500G	Offy	1	20	Finished	2,800
Bob Christie	3	Federal Engineering No.65	Federal Automotive Assoc.	Kurtis 500D	Offy	5	20	Finished	1,500
Bill Cheesbourg	7	Wright-Knight No.25	Ashley Wright	Kurtis 4000D	Offy	8	19	Flagged	300
Bill Randall	12	Safety Auto Glass No.95	Shreve, Perkins, Greismer	Kurtis 500C	Offy	13	10	Oil leak	100
1960									
Trenton, NJ — 100 Mile — 10 Apr									
A.J. Foyt	4	Bowes Seal Fast No.5	Bignotti-Bowes Racing	Kurtis D	Offy	5	100	Finished	1,195
Chuck Arnold	13	Hardwood Door No.88	Ashley Wright	Kurtis 4000D	Offy	20	96	Flagged	338
Al Herman	14	Ray Brady No.59	Ray T. Brady	Kurtis 4000D	Offy	22	92	Flagged	300
Ralph Liguori	16	Federal Engineering No.38	Federal Automotive Assoc.	Kurtis 4000D	Offy	19	62	Wrecked backstretch	225
Don Freeland	21	Vargo No.63	Jake Vargo	Kurtis 4000D	Offy	11	18	Radiator leak	112
George Constantine	99	Tassi Vatis No.94	Tassi Vatis	Kurtis 4000D	Offy		0	DNS, too slow	0
Indianapolis, IN — 500 Mile — 30 May									
Bob Christie	10	Federal Engineering No.38	Federal Automotive Assoc.	Kurtis	Offy	14	200	3:44:51.54 133.416	6,700
Gene Hartley	14	Sumar No.48	Chapman S. Root	Kurtis	Offy	24	196	Flagged	5,710
Shorty Templeman	17	Federal Engineering No.26	Federal Automotive Assoc.	Kurtis	Offy	19	191	Flagged	5,100

Driver	Fin Pos	Car Name	Entrant	Chassis	Engine	Start Pos	Laps	Time/Speed/Out	Purse
Don Freeland	22	Ross Babcock No.73	Racing Associates	Kurtis	Offy	11	129	Magneto	4,310
A.J. Foyt	25	Bowes Seal Fast No.5	Bignotti-Bowes Racing	Kurtis	Offy	16	90	Clutch	4,220
Eddie Russo	26	Go-Kart No.46	Ollie Prather	Kurtis	Offy	29	84	Wrecked SE	4,140
Gene Force	28	Roy McKay No.37	Roy McKay	Kurtis	Offy	20	74	Lost brakes	4,480
Dempsey Wilson	33	Bryant Heating No.23	Duke Donaldson	Kurtis	Offy	33	11	Brkn magneto point	4,380
Jimmy Daywalt	99	Bryant Heating No.23	Duke Donaldson	Kurtis	Offy	26	0	Driven by Wilson	0
Al Keller	99	McKay No.35	Roy McKay	Kurtis	Offy		0	Did not qualify	0
Chuck Hulse	99	Chapman No.43	H.A. Chapman	Kurtis	Offy		0	Wrecked practice	0
Lee Drollinger	99	Ray Brady No.58	Ray T. Brady	Kurtis	Offy		0	Wrecked practice	0
Milwaukee, WI — 100 Mile — 5 Jun									
A.J. Foyt	2	Bowes Seal Fast No.5	Bignotti-Bowes Racing	Kurtis	Offy	4	100	Finished	4,228
Shorty Templeman	13	Federal Engineering No.26	Federal Automotive Assoc.	Kurtis	Offy	9	98	Flagged	559
Bob Veith	14	Roy McKay No.37	Roy McKay	Kurtis	Offy	10	97	Flagged	497
Don Freeland	99	Ross Babcock No.73	Racing Associates	Kurtis	Offy		0	DNS, too slow	0
Eddie Johnson	99	Hardwood Door No.88	Ashley Wright	Kurtis	Offy		0	DNS, too slow	0
Gene Hartley	99	Sumar No.48	Chapman S. Root	Kurtis	Offy		0	DNS, too slow	0
Johnny Tolan	99	Queen Of Them All No.62	Myron Osborne	Kurtis	Offy		0	DNS, too slow	0
Cliff Griffith	99	North Electric No.29	William Tucker	Kurtis	Offy		0	DNS, too slow	0
Dempsey Wilson	99	Bryant Heating No.23	Duke Donaldson	Kurtis	Offy		0	DNS, too slow	0
Bud Tinglestad	99	Wheeler-Foutch No.87	Leroy Foutch	Kurtis	Offy		0	DNS, too slow	0
Norm Hall	99	Roy McKay No.36	Roy McKay	Kurtis	Offy		0	DNS, too slow	0
Bruce Jacobi	99	Safety Auto Glass No.95	Perkins/Griesemer/ Nicholas	Kurtis	Offy		0	DNS, too slow	0
Langhorne, PA — 100 Mile — 19 Jun									
Jack Rounds	7	Hart Fullerton No.52	Hart Fullerton	Kurtis 4000	Offy	4	98	Flagged	824
Bill Homeier	9	Tassi Vatis No.94	Tassi Vatis	Kurtis 4000	Offy	13	96	Flagged	709
Eddie Johnson	99	Hardwood Door No.88	Ashley Wright	Kurtis 4000	Offy		0	DNS, too slow	0
Mike Magill	99	Ray Brady No.89	Ray Brady	Kurtis 4000	Offy		0	DNS, too slow	0
Chuck Arnold	99	Vargo No.63	Jake Vargo	Kurtis 4000	Offy		0	DNS, too slow	0
Springfield, IL — 100 Mile — 20 Aug									
Jack Rounds	11	Hart Fullerton No.52	Hart Fullerton	Kurtis 4000	Offy	12	76	Magneto	532
Bill Homeier	12	Tassi Vatis No.94	Tassi Vatis	Kurtis 4000	Offy	11	52	Blew engine	437
Shorty Templeman	99	Federal Engineering No.26	Federal Automotive Assoc.	Kurtis 4000	Offy		0	DNS, too slow	0
Eddie Johnson	99	Hardwood Door No.88	Ashley Wright	Kurtis 4000	Offy		0	DNS, too slow	0
Chuck Arnold	99	Vargo No.63	Jake Vargo	Kurtis 4000	Offy		0	DNS, too slow	0
Milwaukee, WI — 200 Mile — 28 Aug									
A.J. Foyt	2	Bowes Seal Fast No.5	Bignotti-Bowes Racing	Kurtis	Offy	14	200	Finished	5,455
Shorty Templeman	10	Federal Engineering No.26	Federal Automotive Assoc.	Kurtis	Offy	15	193	Flagged	1,038
Bud Tinglestad	99	Bryant Heating No.23	Duke Donaldson	Kurtis	Offy		0	DNS, too slow	0
Bob McLean	99	Dayton Steel Foundry No.77	George Walther, Jr	Kurtis	Offy		0	DNS, too slow	0
Bill Cheesbourg	99	Hardwood Door No.88	Ashley Wright	Kurtis	Offy		0	DNS, too slow	0
Bill Homeier	99	Tassi Vatis No.94	Tassi Vatis	Kurtis 4000D	Offy		0	DNS, too slow	0
Norm Hall	99	North Electric No.29	William Tucker	Kurtis	Offy		0	DNS, too slow	0
Bruce Jacobi	99	Safety Auto Glass No.95	Perkins/Griesemer /Nicholas	Kurtis	Offy		0	DNS, too slow	0
Wib Spaulding	99	Connor Speed Shop No.95	Edward Connor	Kurtis 3000D	Chrysl	0		Did not qualify	0
DuQuoin, IL — 100 Mile — 5 Sep									
Shorty Templeman	10	Federal Engineering No.26	Federal Automotive Assoc.	Kurtis 4000	Offy	13	53	Flagged	571
Parnelli Jones	98	Federal Engineering No.26	Federal Automotive Assoc.	Kurtis 4000	Offy		41	Rel Templeman	0
Bill Homeier	99	Tassi Vatis No.94	Tassi Vatis	Kurtis 4000	Offy		0	DNS, too slow	0
Jim McWithey	99	Hart Fullerton No.52	Hart Fullerton	Kurtis 4000	Offy		0	DNS, too slow	0
Dick Rathmann	99	Hardwood Door No.88	Ashley Wright	Kurtis 4000	Offy		0	DNS, too slow	0
Jimmy Davies	99	Lutes Truck Parts No.97	W.J. Lutes	Kurtis 4000	Offy		0	DNS, too slow	0
Chuck Arnold	99	Vargo No.63	Jake Vargo	Kurtis 4000	Offy		0	DNS, too slow	0
Syracuse, NY — 100 Mile — 10 Sep									
Shorty Templeman	9	Federal Engineering No.26	Federal Automotive Assoc.	Kurtis 4000	Offy	8	100	Finished	565
Bob Cleberg	99	Tassi Vatis No.94	Tassi Vatis	Kurtis 4000	Offy		0	DNS, too slow	0
Jack Turner	99	Ray Brady No.59	Ray T. Brady	Kurtis 4000	Offy		0	DNS, too slow	0
Indianapolis, IN — 100 Mile — 17 Sep									
Shorty Templeman	9	Federal Engineering No.26	Federal Automotive Assoc.	Kurtis 4000	Offy	7	98	Flagged	1,213
Jack Turner	99	Hart Fullerton No.52	Hart Fullerton	Kurtis 4000	Offy		0	DNS, too slow	0
Jimmy Davies	99	Lutes Truck Parts No.97	W.J. Lutes	Kurtis 4000	Offy		0	DNS, too slow	0
Bob Cleberg	99	Tassi Vatis No.94	Tassi Vatis	Kurtis 4000	Offy		0	DNS, too slow	0
Jim Hemmings	99	Dayton Steel Foundry No.72	George Walther, Jr	Kurtis 4000	Offy		0	DNS, too slow	0
Bud Tinglestad	99	Ray Brady No.59	Ray T. Brady	Kurtis 4000	Offy		0	DNS, too slow	0
Dick Rathmann	99	Hardwood Door No.88	Ashley Wright	Kurtis 4000	Offy		0	DNS, too slow	0
Bruce Jacobi	99	Safety Auto Glass No.95	Perkins/Griesemer Nicholas	Kurtis FE	Offy		0	DNS, too slow	0
Johnny Coy	99	Stoye No.85	Adolph Stoye	Kurtis 2000	Offy		0	DNS, too slow	0

Driver	Fin Pos	Car Name	Entrant	Chassis	Engine	Start Pos	Laps	Time/Speed/Out	Purse
Trenton, NJ — 100 Mile — 25 Sep									
A.J. Foyt	3	Bowes Seal Fast No.5	Bignotti-Bowes Racing	Kurtis	Offy	2	100	Finished	1,980
Bud Tinglestad	14	Bryant Heating No.23	Duke Donaldson	Kurtis	Offy	19	98	Flagged	300
Jim McWithey	15	Federal Engineering No.26	Federal Automotive	Kurtis 4000D	Offy	7	97	Flagged	225
Jim Packard	20	Vargo No.63	Jake Vargo	Kurtis 4000D	Offy	22	73	Mechanical	112
Bob Cleberg	99	Tassi Vatis No.94	Tassi Vatis	Kurtis 4000D	Offy		0	DNS, too slow	0
Bobby Marshman	99	Ray Brady No.59	Ray T. Brady	Kurtis 4000D	Offy		0	DNS, too slow	0
Johnny Coy	99	Stoye No.85	Adolph Stoye	Kurtis 2000D	Offy		0	DNS, too slow	0
Jim Packard	99	Hardwood Door No.88	Ashley Wright	Kurtis	Offy		0	Wrecked	0
Sacramento, CA — 100 Mile — 30 Oct									
Jack Rounds	18	Hart Fullerton No.52	Hart Fullerton	Kurtis 4000	Offy	16	1	Wrecked	137
Chuck Hulse	99	Tassi Tatis No.94	Tassi Vatis	Kurtis 4000	Offy		0	DNS, too slow	0
Bill Hyde	99	Scovell No.65	Robert Scovell	Kurtis 2000	Offy		0	DNS, too slow	0
Mike McGreevy	99	Connor Speed Shop No.91	Edward Connor	Kurtis 3000	Chrysl	0		DNS, too slow	0
Phoenix, AZ — 100 Mile — 20 Nov									
Bill Hyde	15	Scovell No.78	Robert Scovell	Kurtis 2000	Offy	8	13	Poor handling	207
Chuck Hulse	17	Tassi Vatis No.94	Tassi Vatis	Kurtis 4000	Offy	7	4	Wrecked	118
Ernie Koch	98	Scovell No.78	Robert Scovell	Kurtis 2000	Offy		2	Relieved Hyde 14-15	0
Shorty Templeman	99	Scovell No.65	Robert Scovell	Kurtis 2000	Offy		0	DNS, too slow	0
Langhorne, PA — 10 Mile — 13 Apr									
Rodger Ward	99	Ray Brady No.59	Ray T. Brady	Kurtis 4000	Offy		0	DNS, too slow	250
1961									
Trenton, NJ — 100 Mile — 9 Apr									
A.J. Foyt	5	Bowes Seal Fast No.1	Bignotti-Bowes Racing	Kurtis	Offy	5	99	Flagged	1,157
Jiggs Peters	9	Ray Brady No.46	Ray Brady	Kurtis 4000D	Offy	11	97	Flagged	656
A.J. Shepherd	11	Wheeler-Foutch No.87	Leroy Foutch	Kurtis	Offy	21	96	Flagged	548
Chuck Hulse	12	Tassi Vatis No.37	Tassi Vatis	Kurtis 4000D	Offy	18	96	Flagged	450
Red Riegel	99	Vargo No.63	Jake Vargo	Kurtis 4000D	Offy		0	DNS, too slow	0
Bill Randall	99	Safety Auto Glass No.95	Perkins/Griesemer Nicholas	Kurtis			0	DNS, too slow	0
Tony Bonadies	99	Donaldson No.23	Duke Donaldson	Kurtis			0	Did not qualify	0
Tony Romit	99	Stoye No.85	Adolphe Stoye	Kurtis 2000D			0	Did not qualify	0
Indianapolis, IN — 500 Mile — 30 May									
Norm Hall	10	Federal Engineering No.34	Federal Automotive Assoc.	Kurtis 500C	Offy	32	200	3:43:42.39 134.104	8,250
Bob Christie	17	North Electric No.32	William Tucker	Kurtis	Offy	16	132	Burned piston	4,850
Jack Turner	25	Bardahl No.45	Fred Gerhardt	Kurtis	Offy	21	52	Wrecked, MS	5,720
Jimmy Daywalt	31	Schulz Fueling Equipment No.55	C.O. Prather	Kurtis 500G	Offy	23	27	Broken brake line	4,890
Bob Veith	99	Bardahl No.23	Fred Gerhardt	Kurtis	Offy		0	DNS, too slow	0
Norm Hall	99	Concannon Car Co. Flying No.92	Joseph Concannon	Kurtis	Offy		0	DNS, too slow	0
Cotton Farmer	99	Bardahl No.23	Fred Gerhardt	Kurtis	Offy		0	Did not qualify	0
Chuck Hulse	99	Tassi Vatis No.37	Tassi Vatis	Kurtis	Offy		0	Wrecked practice	0
Ronnie Duman	99	Ray Brady No.43	Ray Brady	Kurtis	Offy		0	Did not qualify	0
Dick Rathmann	99	Paxton Products No.75	Andy Granatelli	Kurtis	NoviSc	0		Did not qualify	0
Bob Veith	99	Federal Engineering No.85	Federal Automotive Assoc	Kurtis	Offy		0	Did not qualify	0
Don Davis	99	Federal Engineering No.85	Federal Automotive Assoc	Kurtis	Offy		0	Did not qualify	0
Jack Rounds	99	M.P.C. No.87	Harry Beck	Kurtis	Offy		0	Did not qualify	0
Chuck Weyant	99	Drewry's No.88	Ashley I. Wright	Kurtis	Offy		0	Did not qualify	0
Norm Hall	99	Drewry's No.88	Ashley I. Wright	Kurtis	Offy		0	Did not qualify	0
Bill Randall	99	Safety Auto Glass No.95	Perkins/Griesemer/ Nicholas	Kurtis	Offy		0	Did not qualify	0
Milwaukee, WI — 100 Mile — 4 Jun									
Don Freeland	9	Bardahl No.23	Fred Gerhardt	Kurtis	Offy	13	99	Flagged	987
Chuck Hulse	14	Tassi Vatis No.37	Tassi Vatis	Kurtis	Offy	16	93	Flagged	541
Norm Hall	99	Federal Engineering No.85	Federal Automotive Assoc	Kurtis	Offy		0	DNS, too slow	0
Cliff Spalding	99	Connor's Speed Shop No.90	Edward Connor	Kurtis 3000D	Chrysl	0		DNS, too slow	0
Tony Bonadies	99	Deakin No.91	William R. Deakin	Kurtis	Offy		0	DNS, too slow	0
Langhorne, PA — 100 Mile — 18 Jun									
Red Riegel	7	Vargo No.63	Jake Vargo	Kurtis 4000	Offy	18	94	Flagged	952
Jiggs Peters	12	Ray Brady No.46	Ray Brady	Kurtis 4000	Offy	14	68	Overheated	558
Chuck Hulse	18	Tassi Vatis No.37	Tassi Vatis	Kurtis 4000	Offy	10	5	Broken fuel pump	223
Johnny Coy	99	Stoye No.85	Adolphe Stoye	Kurtis 2000			0	DNS, too slow	0
Milwaukee, WI — 200 Mile — 20 Aug									
Bob Christie	9	North Electric No.32	William Tucker	Kurtis	Offy	10	194	Flagged	1,247
Bud Tinglestad	14	Federal Engineering No.34	Federal Automotive Assoc.	Kurtis D	Offy	26	186	Mechanical	667
Allen Crowe	16	Federal Engineering No.85	Federal Automotive Assoc	Kurtis	Offy	21	181	Mechanical	417
Chuck Hulse	17	Tassi Vatis No.37	Tassi Vatis	Kurtis	Offy	15	173	Mechanical	333
Leon Clum	18	Competition Engineering No.23	Walter Meskowski	Kurtis	Offy	19	173	Mechanical	333

Driver	Fin Pos	Car Name	Entrant	Chassis	Engine	Start Pos	Laps	Time/Speed/Out	Purse
Norm Hall	99	Drewry's No.88	Ashley I. Wright	Kurtis	Offy		0	DNS, too slow	0
Tony Bonadies	99	Deakin No.91	William R. Deakin	Kurtis	Offy		0	DNS, too slow	0
Ray Wearne	99	Connor's Speed Shop No.90	Edward Connor	Kurtis 3000D	Chrysl	0		DNS, too slow	0
Springfield, IL — 100 Mile — 21 Aug									
Chuck Hulse	13	Tassi Vatis No.37	Tassi Vatis	Kurtis 4000	Offy	9	88	Overheated	344
Bud Tinglestad	99	Federal Engineering No.34	Federal Automotive Assoc.	Kurtis 4000	Offy		0	DNS, too slow	0
Red Riegel	99	Vargo No.63	Jake Vargo	Kurtis 4000	Offy		0	DNS, too slow	0
Jiggs Peters	99	Ray Brady No.46	Ray Brady	Kurtis 4000	Offy		0	DNS, too slow	0
DuQuoin, IL — 100 Mile — 4 Sep									
Chuck Hulse	4	Tassi Vatis No.37	Tassi Vatis	Kurtis 4000	Offy	4	100	Finished	1,378
Bud Tinglestad	11	Federal Engineering No.34	Federal Automotive Associates	Kurtis 4000	Offy	15	98	Flagged	601
Syracuse, NY — 100 Mile — 9 Sep									
Len Sutton	9	Tassi Vatis No.37	Tassi Vatis	Kurtis	Offy	15	99	Flagged	565
Bud Tinglestad	99	Federal Engineering No.34	Federal Automotive Associates	Kurtis 4000	Offy		0	DNS, too slow	0
Red Riegel	99	Vargo No.63	Jake Vargo	Kurtis 4000	Offy		0	DNS, too slow	0
Jiggs Peters	99	Ray Brady No.46	Ray Brady	Kurtis 4000	Offy		0	DNS, too slow	0
Vern Harriman	99	Stoye No.85	Adolphe Stoye	Kurtis 2000	Offy		0	DNS, too slow	0
Indianapolis, IN — 100 Mile — 16 Sep									
Bobby Grim	99	Tassi Vatis No.37	Tassi Vatis	Kurtis 4000			0	DNS, too slow	0
Norm Hall	99	Kelly No.88	Ashley I. Wright	Kurtis 4000			0	DNS, too slow	0
Allen Crowe	99	Federal Engineering No.34	Federal Automotive Associates	Kurtis 4000	Offy		0	DNS, too slow	0
Jim McWithey	99	Hart Fullerton No.29	Hart Fullerton	Kurtis 4000	Offy		0	DNS, too slow	0
Vern Harriman	99	Stoye No.85	Adolphe Stoye	Kurtis 2000			0	DNS, too slow	0
Trenton, NJ — 100 Mile — 24 Sep									
Bob Christie	10	North Electric No.32	William Tucker	Kurtis	Offy	20	97	Flagged	663
Bruce Jacobi	99	Tassi Vatis No.37	Tassi Vatis	Kurtis	Offy		0	DNS, too slow	0
Red Riegel	99	Vargo No.63	Jake Vargo	Kurtis D	Offy		0	DNS, too slow	0
Chuck Arnold	99	Federal Engineering No.34	Federal Automotive Assoc..	Kurtis	Offy		0	DNS, too slow	0
Vern Harriman	99	Stoye No.85	Adolphe Stoye	Kurtis 2000D	Offy		0	DNS, too slow	0
Hal Rettberg	99	Ray Brady No.46	Ray Brady	Kurtis 4000D	Offy		0	DNS, too slow	0
Sacramento, CA — 100 Mile — 29 Oct									
Ray Wearne	17	Connor's Speed Shop No.90	Edward Connor	Kurtis 3000	Chrysl	17	10	Black flag, too slow	150
Leroy Neumayer	99	Tassi Vatis No.37	Tassi Vatis	Kurtis 4000			0	DNS, too slow	0
Russ Long	99	C & T Automotive No.38	Clem Tebow	Kurtis 4000	Offy		0	DNS, too slow	0
Bud Sterrett	99	Malcolm Miller No.50	Malcolm Miller	Kurtis 4000	Offy		0	DNS, too slow	0
Phoenix, AZ — 100 Mile — 19 Nov									
Leroy Neumayer	13	Tassi Vatis No.37	Tassi Vatis	Kurtis 4000	Offy	17	84	Flagged	277
Ray Wearne	99	Connor's Speed Shop No.90	Edward Connor	Kurtis 3000	Chrysl	0		Did not qualify	0
1962									
Trenton, NJ — 100 Mile — 8 Apr									
Jim McElreath	16	Ollie Prather No.15	Ollie Prather	Kurtis	Offy	12	61	Engine	224
Chuck Hulse	22	Federal Engineering No.19	Federal Automotive Assoc.	Kurtis 500C	Offy	15	4	Broken torsion bar	75
Ray Brown	99	Vargo No.63	Jake Vargo	Kurtis 4000D	Offy		0	DNS, too slow	0
Leroy Neumayer	99	Albany, NY No.79	Tassi Vatis	Kurtis 4000D	Offy		0	DNS, too slow	0
Jiggs Peters	99	Ray Brady No.43	Ray T. Brady	Kurtis 500G	Offy		0	DNS, too slow	0
Hugh Randall	99	Wheeler-Foutch No.87	Leroy Foutch	Kurtis 500G	Offy		0	Did not qualify	0
Indianapolis, IN — 500 Mile — 30 May									
Jim McElreath	6	Schulz Fueling Equipment No.15	Ollie Prather	Kurtis	Offy	7	200	3:36:22.02 138.653	10,366
Chuck Hulse	21	Federal Engineering No.19	Federal Automotive Assoc.	Kurtis 500C	Offy	16	91	Broken fuel pump	5,531
Jimmy Daywalt	22	Albany, NY No.79	Tassi Vatis	Kurtis	Offy	33	74	Broken transmission	5,351
Jack Turner	29	Bardahl No.45	Fred Gerhardt	Kurtis	Offy	25	17	Wrecked MS	5,141
Bob Christie	30	North Electric No.29	William Tucker, Inc	Kurtis	Offy	31	17	Wrecked MS	5,311
Chuck Stevenson	99	Tropicana No.16	Andy Granatelli	Kurtis	NoviSc	0		Did not qualify	0
Bob Mathouser	99	Ray Brady No.43	Ray T. Brady	Kurtis 500G	Offy		0	Wrecked practice	0
Jimmy Daywalt	99	Goff Brothers Drumstick No.46	Maurice F. Goff	Kurtis	Offy		0	Wrecked practice	0
Mike McGreevy	99	Goff Brothers Drumstick No.46	Maurice F. Goff	Kurtis	Offy		0	Did not qualify	0
Chuck Arnold	99	Goff Brothers Drumstick No.46	Maurice F. Goff	Kurtis	Offy		0	Did not qualify	0
Bill Cheesbourg	99	Hotel Tropicana No.59	Andy Granatelli	Kurtis	NoviSc	0		Did not qualify	0
Al Miller	99	Federal Engineering No.64	Federal Automotive Assoc.	Kurtis 500E	Offy		0	Did not qualify	0
Jim Hemmings	99	Kurtis-Kraft No.65	Roy McKay	Kurtis	Offy		0	Did not qualify	0
Herb Hill	99	Grizzly Brakes No.69	Robert Peterson	Kurtis 500A	Chevy	0		Did not qualify	0
Johnny Coy	99	WRD No.92	William R. Deakin	Kurtis 500G	Offy		0	Did not qualify	0
Jim Hemmings	99	Speed Enterprise No.94	Leonard G. Smith	Kurtis 500G	Offy		0	Did not qualify	0

Driver	Fin Pos	Car Name	Entrant	Chassis	Engine	Start Pos	Laps	Time/Speed/Out	Purse
Milwaukee, WI — 100 Mile — 10 Jun									
Jim McElreath	14	Ollie Prather No.15	Ollie Prather	Kurtis 500G	Offy	14	85	Flagged	566
Chuck Hulse	99	Federal Engineering No.19	Federal Automotive Assoc.	Kurtis 500C	Offy		0	DNS, too slow	0
Hugh Randall	99	Federal Engineering No.64	Federal Automotive Assoc.	Kurtis 500E	Offy		0	DNS, too slow	0
Tommy Copp	99	Ken-ette No.73	Kennette Kelly	Kurtis 4000D	Offy		0	DNS, too slow	0
Mike McGreevy	99	Goff Brothers Drumstick No.46	Maurice F. Goff	Kurtis	Offy		0	DNS, too slow	0
Bob Christie	99	North Electric No.29	William Tucker, Inc	Kurtis	Offy		0	DNS, too slow	0
Herb Hill	99	Grizzly Brakes No.69	Robert Peterson	Kurtis 500A	Chevy	0		DNS, too slow	0
Langhorne, PA — 100 Mile — 1 Jul									
Cotton Farmer	8	Hart Fullerton No.29	Hart Fullerton	Kurtis 400D	Offy	11	97	Flagged	765
Chuck Hulse	13	Albany, NY No.79	Tassi Vatis	Kurtis 4000	Offy	10	93	Flagged	459
Leroy Neumayer	15	Federal Engineering No.19	Federal Automotive Assoc.	Kurtis 4000	Offy	14	72	Hit fence	357
Bob Mathouser	16	Vargo No.63	Jake Vargo	Kurtis 4000	Offy	7	32	Wrecked	306
Elmer George	98	Vargo No.63	Jake Vargo	Kurtis 4000	Offy		17	Rel Mathouser 33-49	0
Hugh Randall	98	Vargo No.63	Jake Vargo	Kurtis 4000	Offy		2	Rel Mathouser 50-51	0
Hugh Randall	99	Ray Brady No.46	Ray T. Brady	Kurtis 4000	Offy		0	DNS, too slow	204
Trenton, NJ — 150 Mile — 22 Jul									
Chuck Hulse	13	Albany, NY No.79	Tassi Vatis	Kurtis 400D	Offy	12	136	Flagged - rain	450
Leroy Neumayer	14	Federal Engineering No.19	Federal Automotive Assoc.	Kurtis 500C	Offy	26	134	Flagged - rain	400
Bob Christie	16	North Electric No.29	William Tucker, Inc	Kurtis	Offy	20	109	Radiator	300
Jim Hemmings	20	WRD No.92	William R. Deakin	Kurtis 500G	Chevy	19	65	Broken value	100
Gig Stephens	99	Ray Brady No.43	Ray T. Brady	Kurtis 500G	Offy		0	DNS, too slow	0
Springfield, IL — 100 Mile — 19 Aug									
Jiggs Peters	11	Albany, NY No.79	Tassi Vatis	Kurtis 4000	Offy	13	96	Flagged	483
Shorty Templeman	13	Federal Engineering No.19	Federal Automotive Assoc.	Kurtis 4000	Offy	18	94	Flagged	395
Cotton Farmer	14	Hart Fullerton No.36	Hart Fullerton	Kurtis 4000	Offy	11	72	Flagged	351
Ralph Liguori	99	Ken-ette No.73	Kennette Kelly	Kurtis 4000	Offy		0	DNS, too slow	0
Al Miller	99	Mitchell Gallas No.95	Mitchell Gallas	Kurtis 4000	Offy		0	DNS, too slow	0
Milwaukee, WI — 200 Mile — 20 Aug									
Bill Cheesbourg	12	North Electric No.29	William Tucker, Inc	Kurtis	Offy	24	187	Flagged	923
Shorty Templeman	18	Federal Engineering No.19	Federal Automotive Assoc.	Kurtis 4000D	Offy	22	50	Magneto	420
Leroy Neumayer	99	Ollie Prather No.15	Ollie Prather	Kurtis 500G	Offy		0	DNS, too slow	0
Jiggs Peters	99	Albany, NY No.79	Tassi Vatis	Kurtis	Offy		0	DNS, too slow	0
Bob Mathouser	99	Federal Engineering No.39	Federal Automotive Assoc.	Kurtis 500C	Offy		0	DNS, too slow	0
Herb Hill	99	Grizzly Brakes No.69	Robert Peterson	Kurtis 500A	Chevy	0		DNS, too slow	0
Langhorne, PA — 100 Mile — 26 Aug									
Cotton Farmer	8	Hart Fullerton No.36	Hart Fullerton	Kurtis 4000	Offy	10	96	Flagged	350
Keith Rachwitz	14	Ray Brady No.43	Ray T. Brady	Kurtis 4000	Offy	17	76	Engine trouble	200
Jiggs Peters	15	Albany, NY No.79	Tassi Vatis	Kurtis 4000	Offy	14	61	Driver fatigue	175
Ralph Liguori	17	Ken-ette No.73	Kennette Kelly	Kurtis 4000	Offy	15	33	Mechanical	125
Syracuse, NY — 100 Mile — 8 Sep									
Al Miller	14	Federal Engineering No.19	Federal Automotive Assoc.	Kurtis 4000	Offy	12	78	Spun out	200
Cotton Farmer	16	Hart Fullerton No.36	Hart Fullerton	Kurtis 4000	Offy	15	69	Worn tire	150
Chuck Hulse	99	Albany, NY No.79	Tassi Vatis	Kurtis 4000	Offy		0	DNS, too slow	0
Carl Williams	99	Mitchell Gallas No.95	Mitchell Gallas	Kurtis 4000	Offy		0	DNS, too slow	0
Jiggs Peters	99	Ray Brady No.43	Ray T. Brady	Kurtis 4000	Offy		0	DNQ, magneto	0
Indianapolis, IN — 100 Mile — 15 Sep									
Bobby Grim	11	Albany, NY No.79	Tassi Vatis	Kurtis 4000	Offy	13	32	Flagged	1,056
Bob Mathouser	98	Albany, NY No.79	Tassi Vatis	Kurtis 4000	Offy		59	Relieved Grim 33-91	0
Jack Turner	99	Hart Fullerton No.36	Hart Fullerton	Kurtis 4000	Offy		0	DNS, too slow	0
Al Miller	99	Federal Engineering No.19	Federal Automotive Assoc.	Kurtis 4000	Offy		0	DNS, too slow	0
Chuck Arnold	99	Mitchell Gallas No.95	Mitchell Gallas	Kurtis 4000	Offy		0	DNS, too slow	0
George Morris	99	Grizzly Brakes No.69	Robert Peterson	Kurtis 500A	Chevy	0		DNS, too slow	0
Trenton, NJ — 200 Mile — 23 Sep									
Bobby Grim	11	Albany, NY No.79	Tassi Vatis	Kurtis	Offy	12	193	Flagged	667
Johnny Rutherford	14	Federal Engineering No.19	Federal Automotive Assoc	Kurtis 4000D	Offy	22	189	Flagged	500
Bill Cheesbourg	17	North Electric No.29	William Tucker, Inc	Kurtis	Offy	16	140	Mechanical	333
Jiggs Peters	24	Ray Brady No.43	Ray T. Brady	Kurtis 4000D	Offy	24	43	Mechanical	111
Sacramento, CA — 100 Mile — 28 Oct									
Art Malone	99	Jiddo/Granatelli No.75	Andy Granatelli	Kurtis 4000	Offy		0	DNS, too slow	0
Ray Wearne	99	Scovell No.69	Robert Scovell	Kurtis 2000	Offy		0	DNS, too slow	0
Buzz Rose	99	Albany, NY No.79	Tassi Vatis	Kurtis 4000	Offy		0	DNS, too slow	0
Phoenix, AZ — 100 Mile — 18 Nov									
Art Malone	18	Jiddo/Granatelli No.75	Andy Granatelli	Kurtis 4000	Offy	5	46	Flagged	156

Driver	Fin Pos	Car Name	Entrant	Chassis	Engine	Start Pos	Laps	Time/Speed/Out	Purse
Howard Osborne	99	Scovell No.69	Robert Scovell	Kurtis 2000	Offy		0	DNS, too slow	0
Bob Mathouser	99	Tassi Vatis No.79	Tassi Vatis	Kurtis 4000	Offy		0	DNS, too slow	0
1963									
Trenton, NJ — 100 Mile — 21 Apr									
Ronnie Duman	9	Federal Engineering No.15	Dan Levine	Kurtis 4000D	Offy	15	97	Flagged	574
Jiggs Peters	14	Vargo Dragway No.48	Jake Vargo	Kurtis 500G	Offy	21	92	Flagged	355
Indianapolis, IN — 500 Mile — 30 May									
Jim Hurtubise	22	Hotel Tropicana No.56	Novi Racing Team, Inc	Kurtis	NoviSc	2	102	Oil leak, black flag	7,400
Art Malone	31	STP No.75	Novi Racing Team, Inc	Kurtis	NoviSc	23	18	Clutch failure	5,150
Bobby Unser	33	Hotel Tropicana No.6	Novi Racing Team, Inc	Kurtis	NoviSc	16	2	Wrecked SW	6,250
Ralph Liguori	99	Schulz Fueling Equipment No.3	Ollie Prather	Kurtis 500G	Offy		0	Did not qualify	0
Norm Hall	99	Federal Engineering No.61	Dan Levine	Kurtis 500C	Offy		0	Did not qualify	0
Junior Johnson	99	American Rubber/Plastic No.73	John Chalik	Kurtis	Offy		0	Did not qualify	0
Milwaukee, WI — 100 Mile — 9 Jun									
Ralph Liguori	16	Schulz Fueling Equipment No.3	Ollie Prather	Kurtis 500G	Offy	18	97	Flagged	462
Bob Christie	99	Federal Engineering No.61	Dan Levine	Kurtis 500C	Offy		0	DNS, too slow	0
Bob Harkey	99	Vargo No.43	Ed Stone	Kurtis 500G	Offy		0	DNS, too slow	0
Langhorne, PA — 100 Mile — 23 Jun									
Bob Mathouser	11	Federal Engineering No.15	Dan Levine	Kurtis 4000	Offy	9	53	Overheated	553
Trenton, NJ — 150 Mile — 28 Jul									
Bob Mathouser	11	Federal Engineering No.15	Dan Levine	Kurtis 4000D	Offy	12	145	Flagged	606
Bruce Jacobi	13	Vargo Dragway No.48	Jake Vargo	Kurtis 500G	Offy	18	142	Flagged	496
Springfield, IL — 100 Mile — 17 Aug									
Bobby Unser	18	Hotel Tropicana No.75	Andy Granatelli	Kurtis 4000	Offy	14	34	Mechanical	189
Jerry Weld	99	Ken-ette No.73	Kenneth Kelly	Kurtis 4000	Offy		0	DNS, too slow	0
Bob McLean	99	Mitchell Gallas No.95	Mitchell Gallas	Kurtis 4000	Offy		0	DNS, too slow	0
Milwaukee, WI — 200 Mile — 18 Aug									
Bob Mathouser	15	Federal Engineering No.61	Dan Levine	Kurtis 500C	Offy	24	189	Flagged	753
Ed Kostenuk	21	City of Victoria No.37	Ed Kostenuk	Kurtis	Offy	18	64	Water hose	188
Bob McLean	99	Mitchell Gallas No.95	Mitchell Gallas	Kurtis 4000D	Offy		0	DNS, too slow	659
Ralph Liguori	99	Concannon No.33	Joe Concannon	Kurtis 500D	Offy		0	DNS, too slow	659
DuQuoin, IL — 100 Mile — 2 Sep									
Jerry Weld	16	Hotel Tropicana No.75	Andy Granatelli	Kurtis 4000	Offy	17	59	Mechanical	335
Lloyd Ruby	17	Ken-ette No.73	Kenneth Kelly	Kurtis 4000	Offy	13	50	Blown engine	279
Bob Mathouser	99	Federal Engineering No.15	Dan Levine	Kurtis 4000	Offy		0	DNS, too slow	0
Bob McLean	99	Mitchell Gallas No.95	Mitchell Gallas	Kurtis 4000	Offy		0	DNS, too slow	0
Indianapolis, IN — 100 Mile — 14 Sep									
Ronnie Duman	9	Federal Engineering No.15	Dan Levine	Kurtis 4000	Offy	13	98	Flagged	1,459
Jerry Weld	99	Hotel Tropicana No.75	Andy Granatelli	Kurtis 4000	Offy		0	DNS, too slow	0
Tommy Copp	99	Mitchell Gallas No.95	Mitchell Gallas	Kurtis 4000	Offy		0	DNS, too slow	0
Trenton, NJ — 200 Mile — 22 Sep									
Ed Kostenuk	11	Kostenuk-Tucker No.37	Ed Kostenuk	Kurtis	Offy	22	187	Flagged	999
Jerry Weld	14	Hotel Tropicana No.75	Andy Granatelli	Kurtis 4000D	Offy	25	151	Flagged	749
George Morris	99	Vargo No.43	Ed Stone	Kurtis 4000D	Offy		0	DNS, too slow	0
Red Amick	99	Mitchell Gallas No.95	Mitchell Gallas	Kurtis 4000D	Offy		0	DNS, too slow	0
Sacramento, CA — 100 Mile — 27 Oct									
Jerry Weld	99	Hotel Tropicana No.75	Andy Granatelli	Kurtis 4000	Offy		0	DNS, too slow	0
Tommy Copp	99	Ed Kostenuk No.79	Ed Kostenuk	Kurtis FE	Offy		0	DNS, too slow	0
Phoenix, AZ — 100 Mile — 17 Nov									
Sonny Helms	12	Vatis No.79	Tassi Vatis	Kurtis 4000	Offy	6	90	Flagged	388
Jerry Weld	13	Hotel Tropicana No.75	Andy Granatelli	Kurtis 4000	Offy	17	90	Flagged	349
1964									
Phoenix, AZ — 100 Mile — 22 Mar									
Johnny Boyd	14	Schulz Refueling No.38	Ollie Prather	Kurtis 500G	Offy	16	97	Flagged	260
Ed Kostenuk	19	Tucker No.37	Ed Kostenuk	Kurtis	Offy	21	65	Spun turn 1	97
Trenton, NJ — 100 Mile — 19 Apr									
Bob Mathouser	9	W.R.D. No.92	William R. Deakin	Kurtis 500G	Offy	20	94	Flagged	622
Ed Kostenuk	12	Tucker No.37	Ed Kostenuk	Kurtis	Offy	21	92	Flagged	479
Bruce Jacobi	14	Vargo Dragway No.43	Jake Vargo	Kurtis 500G	Offy	18	86	Flagged	383

Kurtis-Kraft

Driver	Fin Pos	Car Name	Entrant	Chassis	Engine	Start Pos	Laps	Time/Speed/Out	Purse
Indianapolis, IN — 500 Mile — 30 May									
Bobby Grim	10	Konstant Hot No.16	Tassi Vatis Enterprises	Kurtis	Offy	20	196	Flagged	10,000
Art Malone	11	Studebaker STP No.3	STP Div, Studebaker	Kurtis	NoviSc	30	194	Flagged	9,200
Jim McElreath	21	Studebaker STP No.28	STP Div, Studebaker	Kurtis	NoviSc	26	77	Magneto	5,850
Ralph Liguori	99	Ollie Prather No.38	Ollie Prather	Kurtis 500G	Offy		0	Did not qualify	0
Milwaukee, WI — 100 Mile — 7 Jun									
Bobby Grim	99	Konstant Hot No.16	Tassi Vatis Enterprises	Kurtis	Offy		0	DNS, too slow	25
Langhorne, PA — 100 Mile — 21 Jun									
Dee Jones	16	Vargo Dragway No.43	Jake Vargo	Kurtis 500G	Offy	16	18	Too slow	294
Arnie Knepper	99	Mitch Gallas No.95	Mitchell Gallas	Kurtis 4000	Offy		0	Burned piston pract	196
Trenton, NJ — 150 Mile — 19 Jul									
Bobby Grim	20	Konstant Hot No.16.	Tassi Vatis Enterprises	Kurtis	Offy	18	55	Brkn transmission	141
Chuck Arnold	99	Concannon No.31		Kurtis 500D	Offy		0	DNS, too slow	0
Bob Mathouser	99	W.R.D. No.92	William R. Deakin	Kurtis 500G	Offy		0	Broken camshaft	0
Springfield, IL — 100 Mile — 22 Aug									
Bruce Jacobi	99	Mitch Gallas No.95	Mitchell Gallas	Kurtis 4000	Offy		0	DNS, too slow	0
Milwaukee, WI — 200 Mile — 23 Aug									
Bobby Grim	19	Konstant Hot No.16	Tassi Vatis Enterprises	Kurtis	Offy	20	126	Blew engine	400
Bruce Jacobi	99	Mitch Gallas No.95	Mitchell Gallas	Kurtis 4000D	Offy		0	DNS, too slow	0
DuQuoin, IL — 100 Mile — 7 Sep									
Bob Hurt	99	Mitch Gallas No.95	Mitchell Gallas	Kurtis 4000	Offy		0	DNS, too slow	0
Indianapolis, IN — 100 Mile — 26 Sep									
Hal Rettberg	99	Mitch Gallas No.95	Mitchell Gallas	Kurtis 4000	Offy		0	Did not qualify	0
Trenton, NJ — 200 Mile — 27 Sep									
Chuck Arnold	9	Fred Maday No.31	Fred Maday	Kurtis	Offy	26	189	Flagged	1,286
Bob Mathouser	13	Vargo Dragway No.43	Jake Vargo	Kurtis 500G	Offy	25	128	Flagged	918
Arnie Knepper	23	Konstant Hot No.16	Tassi Vatis Enterprises	Kurtis	Offy	16	54	Spun turn 3	183
Phoenix, AZ — 200 Mile — 22 Nov									
Arnie Knepper	16	Konstant Hot No.16	Tassi Vatis Enterprises	Kurtis	Offy	12	141	Mechanical	479
1965									
Trenton, NJ — 100 Mile — 25 Apr									
Sam Sessions	22	Fred Maday No.38	Fred Maday	Kurtis	Offy	21	0	Did not start	114
Indianapolis, IN — 500 Mile — 31 May									
Arnie Knepper	18	Konstant Hot No.18	Tassi Vatis Enterprises	Kurtis	Offy	19	80	Brkn cylinder wall	10,326
Jim Hurtubise	33	STP Tombstone Life No.59	STP Division Studebaker	Kurtis	NoviSc	23	1	Transmission	8,626
Milwaukee, WI — 100 Mile — 6 Jun									
Arnie Knepper	14	Konstant Hot No.18	Tassi Vatis Enterprises	Kurtis	Offy	13	82	Broken fuel line	744
Dave Paul	99	Paul No.61		Kurtis	Offy		0	DNS, too slow	40
Langhorne, PA — 100 Mile — 20 Jun									
Dave Paul	99	Paul No.61		Kurtis	Offy		0	DNS, too slow	147
Chuck Arnold	99	Nelson No.58		Kurtis	Offy		0	DNS, too slow	147
Trenton, NJ — 150 Mile — 18 Jul									
Chuck Arnold	99	Barzda No.58	Joseph Barzda	Kurtis	Offy		0	Did not start	129
Indianapolis, IN — 150 Mile — 25 Jul									
Bob Mathouser	13	Vargo Dragway No.28	Jake Vargo	Kurtis	Offy	19	77	Flagged	337
Atlanta, GA — 250 Mile — 1 Aug									
Jim Hurtubise	4	STP Tombstone Life No.59	STP Division Studebaker	Kurtis	NoviSc	6	166	Flagged	4,150
Bud Tinglestad	14	STP No.6	STP Division Studebaker	Kurtis	NoviSc	10	160	Flagged	800
Dave Paul	20	Paul No.61		Kurtis	Offy	26	104	Mechanical	760
Dee Jones	24	Midwest Manufacturing No.84		Kurtis D	Offy	29	51	Mechanical	450
Bobby Unser	98	STP No.6	STP Division Studebaker	Kurtis	NoviSc	0		Relieved Tinglestad	0
Langhorne, PA — 125 Mile — 8 Aug									
Wally Dallenbach	13	Nelson No.58	Churchill Harden	Kurtis	Offy	16	116	Flagged	340
Dave Paul	15	Paul No.61		Kurtis	Offy	22	114	Flagged	260
Al Miller	20	Midwest Manufacturing No.84		Kurtis D	Offy	21	41	Mechanical	150
Milwaukee, WI — 150 Mile — 14 Aug									
Wally Dallenbach	99	Nelson No.58		Kurtis	Offy		0	DNS, too slow	186

Driver	Fin Pos	Car Name	Entrant	Chassis	Engine	Start Pos	Laps	Time/Speed/Out	Purse
Dave Paul	99	Paul No.61		Kurtis	Offy		0	DNS, too slow	186
Springfield, IL — 100 Mile — 21 Aug									
Al Miller	12	Midwest Manufacturing No.84		Kurtis	Offy	15	38	Engine	460
Milwaukee, WI — 200 Mile — 22 Aug									
Dave Paul	99	Paul No.61		Kurtis	Offy		0	DNS, too slow	186
DuQuoin, IL — 100 Mile — 6 Sep									
Bob Mathouser	12	W. E. Barnett No.48	W.E. Bennett	Kurtis	Offy	16	61	Broken windshield	566
Al Miller	99	Midwest Manufacturing No.84		Kurtis	Offy		0	DNS, too slow	0
Indianapolis, IN — 100 Mile — 18 Sep									
Al Miller	99	Midwest Manufacturing No.84		Kurtis	Offy		0	DNQ, too slow	0
Trenton, NJ — 200 Mile — 26 Sep									
Billy Foster	24	Konstant Hot No.18	Tassi Vatis Enterprises	Kurtis RE	Offy	15	16	Engine	141
Red Riegel	99	Nelson No.58		Kurtis	Offy		0	Did not qualify	141
Sacramento, CA — 100 Mile — 24 Oct									
Bob Tattersall	17	Midwest Manufacturing No.84		Kurtis	Offy	17	20	Lost oil pressure	394
Gary Congdon	18	Stapp No.15		Kurtis	Offy	13	17	Lost brakes	347
Art Pollard	99	Don Collins No.35		Kurtis	DodgV8	0		Lost oil pressure	0
1966									
Atlanta, GA — 300 Mile — 26 Jun									
Greg Weld	99	STP No.59	Andy Granatelli	Kurtis FE	NoviSc	0		Did not qualify	0
Caldwell	99	Gehlhausen No.84	Carl Gehlhausen	Kurtis 4000D	Offy		0	Did not qualify	0
Indianapolis, IN — 150 Mile — 24 Jul									
Skip Hedrick	99	Foutch No.84	Leroy Foutch	Kurtis FE	Offy		0	DNQ, too slow	0
Langhorne, PA — 150 Mile — 7 Aug									
Chuck Arnold	99	Nelson No.58	Mel Nelson	Kurtis FE	Offy		0	DNS, too slow	0
Springfield, IL — 100 Mile — 20 Aug									
Bob Tattersall	16	Midwest Manufacturing No.84	Carl Gehlhausen	Kurtis 4000	Offy	17	34	Broken oil line	298
DuQuoin, IL — 100 Mile — 5 Sep									
Bob Tattersall	13	Midwest Manufacturing No.84	Carl Gehlhausen	Kurtis 4000	Offy	17	49	Overheated	455
Indianapolis, IN — 100 Mile — 10 Sep									
Bob Tattersall	18	Midwest Manufacturing No.84	Carl Gehlhausen	Kurtis 4000	Offy	15	1	Lost brakes	404
Trenton, NJ — 200 Mile — 25 Sep									
Carl Williams	99	Nelson No.58	Mel Nelson	Kurtis FE	Offy		0	DNS, too slow	0
Sacramento, CA — 100 Mile — 23 Oct									
Bob Tattersall	12	Midwest Manufacturing No.84	Carl Gehlhausen	Kurtis 4000	Offy	8	92	Flagged	470
Ralph Liguori	98	Midwest Manufacturing No.84	Carl Gehlhausen	Kurtis 4000	Offy		0	Relieved Tattersall	0
Phoenix, AZ — 200 Mile — 20 Nov									
Skip Hedrick	99	Foutch No.84	Leroy Foutch	Kurtis FE	Offy		0	DNS, too slow	0
1967									
Milwaukee, WI — 150 Mile — 4 Jun									
Jigger Sirois	99	Cavanaugh No.50		Kurtis FE	Chevy	0		Wrecked	0
Springfield, IL — 100 Mile — 19 Aug									
Bob Wente	7	Midwest Manufacturing No.81	Carl Gehlhausen	Kurtis4000	Offy	15	95	Flagged	828
Jigger Sirois	99	Cavanaugh No.50		Kurtis FE	Chevy	0		DNS, too slow	0
DuQuoin, IL — 100 Mile — 4 Sep									
Bob Wente	99	Midwest Manufacturing No.81	Carl Gehlhausen	Kurtis4000	Offy		0	DNQ, too slow	0
Jigger Sirois	99	Cavanaugh No.50		Kurtis FE	Chevy	0		DNQ, too slow	0
Indianapolis, IN — 100 Mile — 9 Sep									
Jigger Sirois	99	Cavanaugh No.50		Kurtis FE	Chevy	0		Did not qualify	0
Bob Wente	99	Midwest Manufacturing No.81	Carl Gehlhausen	Kurtis4000	Offy		0	Did not qualify	0
Trenton, NJ — 200 Mile — 24 Sep									
Jigger Sirois	99	Cavanaugh No.50		Kurtis FE	Chevy	0		Did not start	0
Sacramento, CA — 100 Mile — 1 Oct									
Wib Spaulding	99	Midwest Manufacturing No.84	Carl Gehlhausen	Kurtis4000	Offy		0	DNQ, too slow	0

Kurtis-Kraft

Driver	Fin Pos	Car Name	Entrant	Chassis	Engine	Start Pos	Laps	Time/Speed/Out	Purse
1968									
Phoenix, AZ — 150 Mile — 7 Apr									
Arnie Knepper	99	Midwest Manufacturing No.84	Carl Gehlhausen	Kurtis4000 D	Offy		0	DNS, too slow	0
Nazareth, PA — 100 Mile — 13 Jul									
Bruce Walkup	12	Midwest Manufacturing No.84	Carl Gehlhausen	Kurtis4000	Offy	14	87	Flagged	461_
Langhorne, PA — 100 Mile — 28 Jul									
Bill Puterbaugh	12	Midwest Manufacturing No.84	Carl Gehlhausen	Kurtis4000 D	Offy	24	90	Flagged	591
Langhorne, PA — 100 Mile — 28 Jul									
Bill Puterbaugh	9	Midwest Manufacturing No.84	Carl Gehlhausen	Kurtis4000 D	Offy	12	87	Flagged	281
Springfield, IL — 100 Mile — 17 Aug									281
Bill Puterbaugh	6	Midwest Manufacturing No.84	Carl Gehlhausen	Kurtis4000	Offy	10	98	Flagged	281
Jigger Sirois	16	Cavanaugh V8 No.108		Kurtis FE	Chevy	17	18	Out	281
DuQuoin, IL — 100 Mile — 2 Sep									281
Tom Bigelow	6	Midwest Manufacturing No.84	Carl Gehlhausen	Kurtis4000	Offy	12	98	Flagged	281
Karl Busson	99	Cavanaugh No.108		Kurtis FE	Chevy		0	DNS, too slow	281
Indianapolis, IN — 100 Mile — 7 Sep									281
Karl Busson	99	Cavanaugh No.108		Kurtis FE	Chevy		0	DNS, too slow	281
Tom Bigelow	99	Midwest Manufacturing No.84	Carl Gehlhausen	Kurtis4000	Offy		0	Did not qualify	281
Sacramento, CA — 100 Mile — 29 Sep									281
Tom Bigelow	14	Midwest Manufacturing No.84	Carl Gehlhausen	Kurtis4000	Offy	14	39	Out	281
Phoenix, AZ - 200 Mile - 17 Nov									281
Karl Busson	99	Cavanaugh No.108		Kurtis FE	Chevy		0	Brkn timing chain	281
1969									281
Nazareth, PA — 100 Mile — 12 Jul									281
Tom Bigelow	7	Midwest Manufacturing No.84	Carl Gehlhausen	Kurtis	Offy	10	83	Flagged	281
Trenton, NJ — 200 Mile — 19 Jul									281
Charles Masters	17	Midwest Manufacturing No.84	Carl Gehlhausen	Kurtis D	Offy	23	67	Flagged	281
Springfield, IL — 100 Mile — 18 Aug									281
Tom Bigelow	99	Midwest Manufacturing No.84	Carl Gehlhausen	Kurtis	Offy		0	DNS, too slow	281
DuQuoin, IL — 100 Mile — 1 Sep									281
Tom Bigelow	8	Midwest Manufacturing No.84	Carl Gehlhausen	Kurtis	Offy	17	99	Flagged	281
Indianapolis, IN — 100 Mile — 6 Sep									281
Tom Bigelow	99	Midwest Manufacturing No.84	Carl Gehlhausen	Kurtis	Offy		0	DNS, too slow	281
Trenton, NJ — 300 Mile — 21 Sep									281
Charles Masters	17	Midwest Manufacturing No.84	Carl Gehlhausen	Kurtis D	Offy	28	72	Ignition	281
1970									281
Springfield, IL — 100 Mile — 22 Aug									281
Tom Bigelow	7	Midwest Manufacturing No.84	Carl Gehlhausen	Kurtis	Offy	12	82	Flagged	281
DuQuoin, IL — 100 Mile — 7 Sep									281
Tom Bigelow	7	Midwest Manufacturing No.84	Carl Gehlhausen	Kurtis	Offy	9	97	Flagged	281
Indianapolis, IN — 100 Mile — 12 Sep									281
Tom Bigelow	99	Midwest Manufacturing No.84	Carl Gehlhausen	Kurtis	Offy		0	Bured piston	281
Sedalia, MO — 100 Mile — 19 Sep									281
Tom Bigelow	6	Midwest Manufacturing No.84	Carl Gehlhausen	Kurtis	Offy	12	97	Flagged	281
Sacramento, CA — 100 Mile — 4 Oct									281
Tom Bigelow	17	Midwest Manufacturing No.84	Carl Gehlhausen	Kurtis	Offy	14	33	Piston	281

Index

Models

Index (continued)

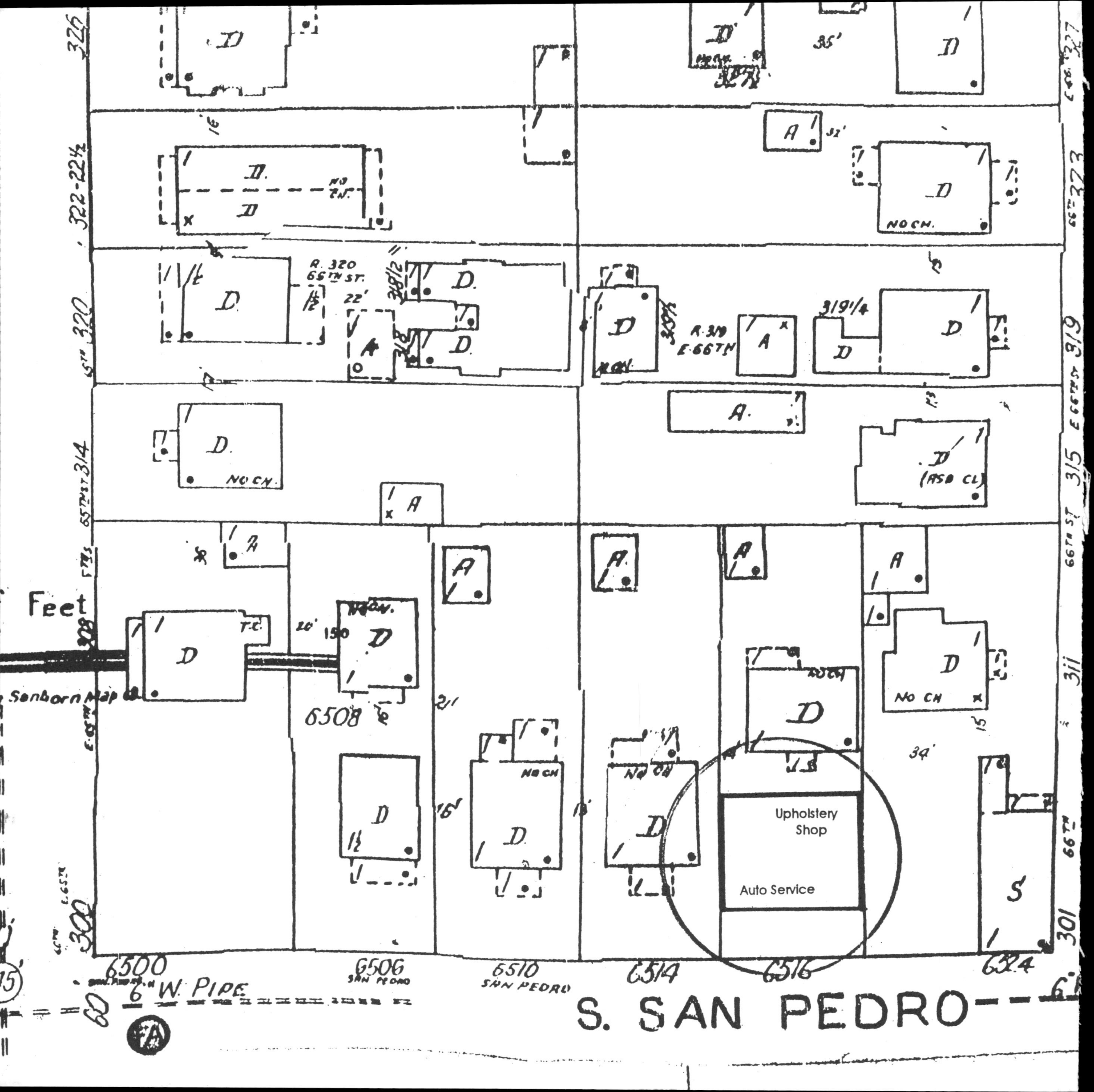
S. SAN PEDRO
Upholstery Shop
Auto Service
6500
6506 San Pedro
6508
6510 San Pedro
6514
6516
6524
6" W. PIPE
Feet
Sanborn Map Co.
300
301
311
314
315
319
320
322-224½
326
No CH.